Forgotten Sea Serpents

by

Malcolm Smith

Forgotten Sea Serpents

Independently published 2020

Copyright © Malcolm Smith, 2020

This book is copyright. Apart from fair dealing for the purposes of private study, research, criticism or review, as permitted under the Copyright Act, no part may be reproduced by any process without written permission.

1. Sea Serpents   2. Cryptozoology   3. Folklore.
4. Marine Mysteries

ISBN: 9798627524566

Cover illustration: contemporary (and probably highly imaginative) drawing of the 1891 Pablo Beach sea serpent.

Chapters

Introduction …**1**

1. Mid-Nineteenth Century (1834 to 1869)…**5**

2. The 1870s and '80s…**16**

3. Strange Visitors to New Zealand…**39**

4. The Last (Forgotten) Sea Serpents of the 19th Century...**48**

5. The *Tresco* Sea Serpent of 1903….**59**

6. Edwardian Sea Serpents...**75**

7. 1914 and After….**95**

8. Lakes and Swamps…**113**

Help Wanted…**123**

Index…**124**

Other Books by the Author…**128**

Introduction

Yes, Virginia, there are such things as sea serpents, although you don't hear so much about them these days - a fact that is probably related more to the behaviour of witnesses than the animals themselves. People are now happy to report, and newspapers to publish, alleged sightings of lake monsters, but sea serpents are no longer respectable.

It wasn't always so. The heyday of the sea serpent was the nineteenth century, when science took it seriously, as did the popular press. Also, during the age of sail, ships moved silently across the surface, and often strayed dramatically from the shipping lanes. Nowadays, huge vessels stick tightly to the appropriate lanes, with their engines producing enough noise to discourage the appearance of any self-respecting sea monster.

Nevertheless, I know from my own research that the Australian press in particular was happy to print stories of sea serpents right up to the time of the Second World War, after which the subject apparently dropped out of their world view. Just the same, sightings continued to be reported by brave souls unafraid of public ridicule.

Over the years, three comprehensive books have been published on the subject. The first was *The Great Sea-Serpent* by Dr. A. C. Oudemans in 1892, followed in 1930 by the retired naval captain, R. T. Gould with *The Case for the Sea-Serpent*. Finally, in 1968, Dr. Bernard Heuvelmans came out with *In the Wake of the Sea-Serpents*. After a lapse of half a century, we may well feel that it is time for a new collection, but at least one must admit that Heuvelmans' study was up to date. His last three reports dated from 1966. Not only that, but for the whole of the 1950s and '60s he was able to record several sightings for each year. Clearly, the phenomena showed no sign of abating. Not only that, but his work was exhaustive; over a period of three centuries, he was able to list 587 sightings. After culling out vague reports, obvious hoaxes, and probable misidentifications, he was still able to keep the score at 358, or more than one a year. No doubt a more rigorous sceptic would be able to reduce it further, but it is nevertheless clear that something very strange is going on. Also, as this book will reveal, there were quite a few which he missed.

By this time, you may well be asking: what exactly do sea serpents look like? In short, they are elongated, and of a comparable length to a whale, but slimmer. They are probably a lot rarer, and their habits are certainly solitary. They do not migrate in pods or herds like some species of whales. Also, he made the point that an elongated animal can wriggle out of the shallows, and so they do not get stranded like whales, and they probably use echolocation, which would prevent their falling foul of our nets.

Apart from that, they exist in a bewildering variety of forms. The first two authors assumed there was only a single species, but Heuvelmans could see this was not the case. While I am not so eager to confirm his assessment of *nine* different types, there are certainly more than one. One type is represented by a short neck and a long series of humps - a "string of buoys", as it is often described. While a snake, or an elongated fish, swims by means of horizontal undulations, the undulations of the "many humped" sea serpent are vertical, indicating it must be a mammal - probably some sort of whale.

Another type presents as a long neck rising out of the water like a periscope. Frequently, the rest of the body can be seen as a short series of large humps trailing behind. To make matters more interesting, sometimes the head sports a pair of large eyes, with occasionally a mane visible on the neck, while at other times no eyes are visible, even close up. Heuvelmans believed they represented two separate species. I am not so sure, but I am unable to hazard a guess as to their identities except to say that I find the two most common speculations - that they are plesiosaurs, or long necked seals - unconvincing, for reasons given in my earlier book, *Australian Sea Serpents*.

In any case, you must bear this variety in mind when you peruse the rest of the book. And, of course, one cannot rule out something like a gigantic eel. The sea still holds many mysteries.

But now we live in the internet age, and one of its development is Trove[1], the project of the Australian National Library to digitalise most Australian newspapers. So far, the major publications, and a great many of the minor ones, up to the mid-1950s, have been put online. It therefore occurred to me that I could use it to locate the original source of sightings I had known only from secondary sources. It worked only too well. By simply doing a search on the phrase, "sea serpent" year by year, I soon uncovered enough new material to more than double the tally of

---

[1] http://trove.nla.gov.au

reported sightings, as documented in my recent book, *Australian Sea Serpents*.

Not only that, I uncovered a whole trove of *foreign* sea serpent sightings. These reports came about in two ways. The first was that the ship's first port of call was in Australia - most often Melbourne - and the captain immediately informed the press of what they had seen perhaps on the opposite side of the world. The second, more common method was for a newspaper - it may have been simply a restricted rural journal, rather than a capital city daily - utilising what served as "the cloud" in the pre-internet era, picked up a story from some, often obscure, foreign newspaper and decided to print it. After that, for the next couple of weeks or couple of months, it would do the rounds of the Australian press. Furthermore, by cross-checking the stories against the accounts in Heuvelmans, I discovered that a lot of these reports were new ie they had been completely missed by all previous researchers.

Now, there happens to be a small but significant group of people who are vitally interested in cryptozoology, the study of unknown animals, in general, and sea serpents in particular. It is for these that I have produced the current volume. Its aim is to provide the original sources for these forgotten sea serpents culled from the Australian press, in the hope that they will render the whole corpus of evidence more comprehensive. However, a few things need to be said about the reports.

Firstly, I have generally cited the first appearance of the story in the Australian press, and have provided both the title and the page number, so that interested parties can check it for themselves. Note, however, that those accounts picked up from foreign newspapers typically do not provide the date of the original report. It may have been a couple of weeks, or even a couple of months, prior.

Secondly, all measurements are given in imperial units. I have therefore converted these into metric in square brackets. I have not, however, attempted to be more precise than the original estimates. Thus, six feet is cited as 1.8 metres, not the correct 1.83 metres. Nobody, looking at the neck of a mysterious monster, is going to be that precise! Similarly, I not converted yards into metres. As a rough estimate of distance, a yard is equivalent to a metre.

Thirdly, as a general rule, although the journalists apparently took the accounts seriously, they were content to report merely the details volunteered by the witnesses. In hardly any case did they bother to ask questions which might have assisted in identifying the animal, or at least the group of animals to which it belonged.

Finally, I haven't included every report. Some were so outrageous, they were clearly hoaxes. Such, for example, are tales about the sea serpent being killed, and its body preserved for scientific study. The format of other accounts suggested they were intended as short stories, rather than factual reports. I have left them out as well. Even so, many of those I have included sound somewhat suspicious, but when in doubt I have felt it is better to allow the reader to make up his own mind.

So don't believe everything you read in this book. But don't disbelieve everything, either. Examine them with an attitude of open minded scepticism. As I said in my previous book, how often does something weird have to be reported before we accept that something weird is going on?

Chapter 1.

Mid-Nineteenth Century (1834 to 1869)

Massachusetts, ?1834.

Sea serpent sightings were very common off the coast of New England in this period. This particular article was taken from the *Launceston Advertiser* (Tasmania) of Monday 5 May 1834, on page 2, citing the *Boston Centinel* in the first instance. However, there is something not quite right, because it states that the event took place on Friday *July* 10th. It could not have referred to the previous year, because that date was a Wednesday in 1833. I suspect that the correct date was Monday, March 10th., 1834. Perhaps my Boston readers can look up an old copy of their newspaper and discover the truth. In any case, the Tasmanian article reads as follows:

SEA SERPENT
From the Mercantile Advertiser and New York Advocate.

A party of 80 or 100 ladies and gentlemen embarked on Friday morning, July 10th, in the steamer Connecticut, for the purpose of taking an excursion in the lower harbour, with the expectation of getting a view of His Serpentine Majesty. About 12 o'clock, when the steam boat was half way between Nahant and the Graves, the monster was seen approaching. A number of gentlemen took the small boat and made directly for it, but unfortunately did not run upon the animal as was intended, owing to a little mismanagement in rowing. The Serpent came within an oar's length of the boat, and without appearing at all alarmed or uneasy took a slight curve towards the steam boat, passed under her stern within fifty or sixty feet [*15 or 18 metres*], and then disappeared. We understand it was the opinion of those in the small boat that he might easily have been struck, but unfortunately there was no harpoon on board. At this time his motion was not undulating, as has sometimes been stated, but rather like the movement of an eel or common water snake. It has been reported that there have been three or more of these strange creatures seen lately, one of which is supposed to be 150 feet long [*45.7 m*]. The one seen yesterday was from 60 to 70 feet [*18 to 21 m*] in length. We would recommend some of our sporting friends, who are skilled in the management of a whale boat, and use of the harpoon, to make an attempt upon the liberty of this marine monster, and there is but little doubt he may be taken. The foregoing account is furnished by a gentleman who was one of the passengers, and had a good opportunity to see the serpent from the small boat, and whose certificate is annexed. This statement in its material bearings is also corroborated by several other gentlemen with whom we have conversed, who were on board the steamer. The excursion of yesterday has afforded a much better opportunity of seeing this strange animal than has occurred for years, and it is not inconsistent with the whole tenor of the statements that have been made at different times by a great number of persons for the last 15 years, since a monster of this description was first announced in our waters. It is admitted on all hands that the appearance of a marine animal of this description would be a most extraordinary occurrence.

But it may be said as an offset, that it would be still more extraordinary, if so many witnesses should be so grossly deceived, as would be the case if no such animal had appeared. One or the other of these extraordinary difficulties is presented for the belief of the public, and we are of opinion that it would not require so great a stretch of credulity to believe in the existence of such an enormous Sea Serpent, as to believe that so many persons could be so greatly deceived. We learn that a gentleman fired at him with a musket from the steamer, but without effect. The shot was given before he had approached so near the steamer as he did a few minutes afterwards. The first thing that attracted the attention of those who were in the steamer, was a peculiar appearance in the water at a distance, supposed to be occasioned by a shoal of small fish that he was apparently pursuing. Three distinct appearances of this kind were observed at the same time afar off, and the steamer made for one of them in pursuit of which the serpent appeared to be. It is therefore inferred by some of the passengers that there are three of the strange animals, as we have before stated. — *Boston Centinel*.

You might notice a distinct lack of detail as to the physical appearance of the animal, other than its size.

Barque *Inconstant*, south east of Africa, 1849

This report was originally published in the *South Australian Register* of Saturday 9 June 1849, on page 4, and was taken up by a large number of other newspapers the following month (!). For your information, the same edition reported that the *Inconstant* had arrived at Adelaide on 8 June, carrying 209 Irish female orphans.

Our readers will remember that some months ago the officers and crew of her Majesty's ship *Daedalus* reported having seen what they called a sea serpent in the latitude of the Cape. But from the description given, Professor Owen has declared his belief the animal seen was a large seal lion, which had probably been carried away by an iceberg. It is a somewhat singular coincidence that a marine animal was seen by the officers and some of the men of the *Inconstant* on the passage to this colony, and not any considerable distance from the spot rendered remarkable by the *Daedalus*'s discovery. We subjoin [sic] a description kindly furnished by Charles Watkins, Esq., Surgeon Superintendent. We hope the writer will lose no time in sending a copy of his drawing and description to some scientific friend in England:-

"On May 1st, 1849, in latitude 38°8 south, longitude 36° 20 east, at half-past nine a.m., my attention was called by the second mate to a strange looking object in the water on the lee bow. Its first appearance was that of a log of wood or large tree; but as the ship neared the object, when at a distance of fifty yards, more or less, we perceived it to be a species of serpent, or something closely resembling one. Its color was of a brownish black. It was going slowly through the water, carrying its head at an angle to the waves. The head resembled that of an alligator. It was covered with scales from the tip of the snout to the end of the neck: the rest of the creature was under water, and apparently about thirty or forty feet long [9-12 m] (no end was seen.) The size round the neck was about the dimensions of a puncheon. While looking at it it sank gradually, and appeared again on the larboard quarter. We soon lost sight of it, as we were going six knots [11 kph]. It was proceeding in the direction E.N.E. It was seen by the master, second mate, and William Vernon, an able seaman, who was at the wheel. These parties have been many years at sea, and have never seen anything like it before. They agree in saying that the drawing is an exact resemblance of the thing seen.

CHARLES WATKINS.
"Barque *Inconstant*."

It would have been nice if the drawing had been included.

Alpha, 1849

As far as I can establish, the original report appeared in the *Melbourne Daily News* of Tuesday 3 July 1849, on page 2.

THE GREAT SEA SERPENT. — The following is an extract from the private log of Captain Edwards, of the "Alpha," furnishing additional particulars, in reference to this monster. "Wednesday, May 30th, p.m. strong breezes at N N.W., and a sharp sea on; about 1.15, I felt a strange shaking of the ship, as if it proceeded from some submarine volcanic eruption, or as if the ship were drawn over a shingle bottom in smooth water. Mr. Thompson, my chief-officer, Mr. George Park, civil engineer, cabin passenger, on board, run on deck as well as myself, when to our infinite surprise, we beheld immediately under our lee quarter, a monster of huge dimensions, three times larger than any whale I had ever seen; it did not partake of the shape of a whale, as it had no fins or broad tail as whales have, neither did it jet up a column of water as whales do when blowing. I was afraid it would have made another rush at the ship, and of course the consequences might have been serious. As to what the animal might have been, I cannot pretend to say - but I never before saw a similar one. It was of a light fawn colour, spotted over behind the shoulders with large brown spots, the head was pointed like that of a porpoise, it had large glassy eyes, the appearance of the shoulder was much darker than the rest of the body, which was the thickest part of it (say 20 feet [*6 m*] in diameter,) from thence diminishing to the tail, to about the size of our mainyard in the slings (say 24 inches [*60 cm*] diameter) and rounded off like that of a worm or caterpillar. His speed seemed to be very great, as after having grated our bottom, he took a turn round, it would seem, as we afterwards saw him astern, and he went away then in a S.E. by S. direction, at about 30 miles an hour [*50 kph*]."

This report did turn up in Heuvelmans' book, and I agree with him that it was probably some huge ray. My dispute is his assumption that it took place in the "waters of the Indian Ocean south of Australia". He was citing a secondary source. However, the newspaper shipping reports indicate that the *Alpha* arrived in Melbourne on 30 June, so the sighting occurred one month prior to arrival. This was the age of sail, and you will remember that the *Inconstant*'s encounter took place 39 days before its arrival in Adelaide. I would suggest, therefore that the *Alpha* met its "monster" half way across the Indian Ocean, but south of the latitude of southern Australia.

British Isles, 1850

In this case, we know exactly where the sighting took place, but not when. In fact, the Australian newspaper appears to have picked up two separate accounts. In any case, the reference is the *South Australian Register* (Adelaide), Wednesday 4 December 1859, on page 4.

THE SEA-SERPENT AGAIN

This mysterious monster of the deep has been seen twice recently - once near the entrance of the British Channel, and subsequently in Dublin Bay; in the first instance by a vessel bound to London from the Isle of France. The following extract from the log of the vessel, the *Lucille*, Captain Benson, furnishes some interesting and authenticated particulars relating to this strange visitant; and it is worthy of remark that all accounts of it, from that of the *Daedalus* down to those supplied by the *Freeman's Journal* of is appearances on the Irish coast, accord closely in their descriptions of this "*monstrum horrendum informe ingens.*" [*huge, horrendous, misshapen monster*]

Abstract from the log of the brig *Lucille*, on her passage from Mauritius for London:-

"In lat. 48.51 N., and long. 12.14, W., observed something extraordinary on the starboard quarter, about a mile [*1.6 km*] distant, coming towards the ship at a brisk rate, and on nearing us found it to be an enormous serpent; which, when abreast of the brig, we judged to be (from our own length, 81 feet [*24.7 m*]) upwards of 100 feet [*30 m*], that length being plainly visible from the deck with our glasses. It had a tremendous flat head, and apparently a horn or fin behind, with large bunches of hair about it; it propelled itself by an undulating snake like motion, and held on a steady course to E.SE. at a rate of 6 or 7 miles per hour [*10 -11 kph*], causing great motion in the water, and leaving a large wake behind it. This strange monster passed within a quarter of a mile [*400 metres*] of the ship, and we watched it, as long as it was visible, from the mainyard with our glasses.

"H. B. Benson, Commander.
"S. E. Suncombe, Second Mate.
"A. W. Owen, late Commander)
   barque *Despatch*, of London) Passengers
"S. G. Reay   )

The reference to Mr. Owen as the "late" Commander does not, of course, mean that he had passed away before the log was published, but merely that he had been *lately* a Commander. Now, falsifying a log is a serious offence, and it is unlikely that four people would be guilty of such a thing for the trivial purpose of a harmless joke on the public. It is noted, too, that they were using telescopes, so the distance involved is not so problematical. I am always wary of reports of flat heads, because that it a feature of real snakes, and the sort of thing hoaxers catch onto. However, the horn or fin behind and the large bunches of hair sound less like a hoaxer's invention. I wonder if the "snake-like" motion meant lateral, rather than vertical, undulations. I suspect not.

The article continues with an account apparently taken from the *Freeman's Journal*.

We subjoin to this the account of its subsequent appearance in Dublin, taken from the source already adverted to:-
"On the 15th of August, Mr Walsh, of Sackville-street, Mr Hogan of Sutton, and several other gentlemen, while enjoying a sail in the yacht of Mr Hogan, had the additional and exciting pleasure of witnessing the evolution of an enormous sea-monster, which more resembled in shape and size the great sea-serpent than any other living being which any of the gentlemen had ever before seen or heard described. Mr Hogan's yacht was, at the time the monster appeared in view (half past 6 p.m.), sailing between Dalkey and Sutton. One of the gentlemen on board the yacht saw the monster at a distance of about half a mile [*800 metres*] rushing with great impetuosity in a direction towards Howth Point. He immediately directed the attention of his companions to the strange visitor, and the whole party continued for several minutes to watch his movements and scrutinise his shape and dimensions. Several portions of the back were in view over the water, and

seemed to resemble 'the coils of a serpent,' to adopt the phraseology of one of the gentlemen who waited on us to describe the circumstances. The head was shaped not unlike that of an eel, and was borne aloft several feet out of the water. The speed at which he moved through the water was estimated at twenty miles an hour [*32 kph*], and he left a wake such as might be expected from a ship of several hundred tons. The gentlemen who saw this monster computed his length at 100 feet [*30 m*]; and Mr Walsh informs us that Mr Hogan, who had been many years at sea, was quite satisfied that the monster was not of the whale tribe, or was not of a species heretofore known to mariners, and described by naturalists."

All the places involved are parts of Dublin. In assessing this account, of course, we must remember the distance involved. Just the same, the description was similar to that of many other sea serpents throughout the world: a head held vertically, and a series of humps behind. The latter suggests that the witnesses were prepared to describe vertical humps or undulations as serpentine. Whether, of course, it was the same animal seen by the *Lucille* at any unreported date prior to that, is anyone's guess.

New Jersey, 1869

New England had a long tradition of sea serpents. Who knows exactly when the following occurred, but it was published in the *Northern Argus* (Rockhampton, Qld) on Monday 8 November 1869, on page 3? It was apparently taken from the English press.

THE SEA SERPENT AGAIN. — The *Newark* (New Jersey) *Courier* has a sea serpent story. — "When to his great surprise and terror, the head of a monster as large as a flour barrel, and having something of the appearance of a dog's head, appeared above the water. It stretched away along the surface, and a black scaly back lifted itself gradually from the water until it appeared, according to Mr. Andrews, twice the length of an ordinary schooner. It swam easily, and with but little motion, occasionally rising its head three or four feet [*say a metre*] above the surface with that peculiar sinuosity common to the snake tribe. Suddenly, with a tremendous splashing, it disappeared from sight, leaving behind a large sea of seething foam. Mr. Andrews acknowledges himself to have been 'scared almost to death' at the sight, and about came to the conclusion, so he says, that he was to be eaten alive, indeed his presence of mind so far forsook him that he dropped both oars, and had some difficulty in recovering them. Having secured them, however, by means of a small paddle which fortunately remained in the bottom of the boat, he undertook to row across the bay; but he had proceeded but a short distance when a terrible splashing from behind caused him to turn round, and there, as he solemnly asserts, within a dozen yards of him, was the head of the monster high up above the surface; and, to add all the more to his terror, it opened its hideous jaws and darted a forked tongue directly at him. To employ the language used by Andrews himself, 'the next thing he knew, he didn't know anything,!' meaning thereby that his terror was so great he apparently lost consciousness. That was the last he saw of the sea serpent."

The problem I have with this, apart from the dramatic nature of the tale, is the reference to a scaly skin and a forked tongue. Hoaxers normally think of the sea serpent as a genuine snake, but all the evidence suggests that it is not so. Although scales have occasionally been reported in otherwise plausible accounts, to my knowledge, a forked tongue, which is the mark of a snake, never has. Indeed, I am not certain the tongue has ever been reported elsewhere. However, I shall leave the plausibility of the story up to you.

New England, 1869

The coasts of New England were famous for sea serpents in the late nineteenth century. They tended to be of the long-necked type, but this one is very unusual. For that reason, I suspect it is genuine. A hoaxer would be more likely to build on the more familiar theme, but go over the top with it (and I have read some which qualify). It was apparently originally published in the *New York Sun* of 30 November 1869, but it was not for another 2½ months that an Australian newspaper picked it up. This is from the *Sydney Mail* of Saturday 12 February 1870, page 15.

THE SEA SERPENT AGAIN
THE MONSTER IN THE GULF STREAM WITH A RECRUIT

The Sea Serpent still lives, and has an heir for the perpetuation of his race. Captain Allen, of the ship Scottish Bride, which arrived at New York on Sunday, brings the latest intelligence from his marine snakeship, the captain having encountered the monster on the 23rd instant [*ie 23 Nov 1869*], in latitude 38°16', longitude 74°08'. The remarkable feature of the meeting was that the old, familiar serpent, fifteen or twenty feet [*4½ to 6 metres*] long, and as big around as a hogshead, was accompanied by a juvenile monster of the same species, only five feet [*1½ metres*] in length. This meeting, as will be seen by reference to the charts, was on the edge of the Gulf Stream, about 200 miles [*320 km*] off Delaware Bay, but as Captain Allen is a credible witness - favourably known by the shipping merchants of New York, and everywhere credited to be an intelligent man, his own narrative of the singular meeting will be read with greater interest than any more studied account :

— Captain Allen is a thorough type of an American skipper, sharp, shrewd, bluff and honest, and has followed the ocean from boyhood, rising by his own energy and merit from a cabin boy to the command of one of the finest clippers sailing from this port.

Captain Allen says that on the 23rd of this month he descended to his cabin after a fruitless effort to get a meridian observation, the sky being too much overcast. He was just about eating his dinner when his second mate descended the cabin stairs, and, in an excited manner, told him his presence was required on deck. Thinking the ship had sprung a leak, or that some other dire mishap had befallen them, he dropped the tempting morsel before him and rushed up. When he arrived on deck he found the crew assembled on the starboard side of the vessel, looking with awe-stricken countenances into the water. Not knowing the meaning of their strange conduct, he also went to the ship's side, and a sight met his eyes the memory of which will never fade.

The weather had been thick and nasty all the morning, the heavens heavily overcast, threatening to pour forth a deluge at any moment, and the wind blowing from all quarters at once. But now there was a dead calm, and the surface of the sea was undisturbed by a ripple. On approaching the side of the vessel, the captain saw in the water beneath a monster such as he had never seen before. It was about 25 feet [*7½ m*] in length, and proportionately thick. Its head was very large and flat, while at each side, on the extreme edge, were set two bright, scintillating eyes, which he says, looked dangerous and wicked. Its back was covered with huge scales, like the crocodile, about three inches [*7½ cm*] in length, which hooked together and formed an impenetrable armour. Its belly was of a tawny yellow colour, and altogether hideous. It was accompanied by a smaller specimen of its own species, and may have been its offspring. This was but a few feet in length, but in shape and colour closely resembled the larger one.

All the efforts of the captain to have the sailors to make some attempt to capture it were abortive. They looked upon it as something supernatural, and were not disposed to meddle with it. The thing was about four feet [*1.2 m*] from the vessel, was lying but a few feet below the surface of the water, and was discernible to all on board. The captain gave orders to have a boat lowered to attack the monster, but in the meantime the attention of the smaller one was called to the presence of the vessel. It raised its head a few inches above the surface, and then went to its larger friend, and seemed to tell it of the circumstance ; but whatever transpired between them, the larger one raised its head, as though to investigate its surroundings, and then, with an easy motion, it dropped into the ocean. In disappearing, it went head downward, and its body described a circle like a hook, its tail raising out of the water, which, the captain says, tapered off to a sharp point.

The calm that had beset the vessel in the morning now gave way to a strong northwest breeze, that as night closed around, burst into a storm, accompanied by vivid lightning and rolling thunder. The ship was tossed about by the waves, which ever and anon broke over her with relentless fury, and during the whole of this fearful night the sailors would not go on deck without lanterns, such was their fear of meeting the monster. Now and then they would go to the captain and ask his opinion on the probability of that occurrence ; but he being no wiser than themselves, would laugh at their fears and bid them go to their work. About morning the storm died away, but until the following day, when they came in sight of land, the brave men entertained an unexpressed dread of the reappearance of the monster.

Captain Allen thinks that the monster came from the regions of Florida, where he has often heard of similar creatures from other shipmasters, and by following the warm current of the Gulf Stream it reached the position where he found it. In his opinion, it is a deep water animal; and he accounts for its appearance so near the surface by the fact of the dark day, and the monster not knowing now high up it was.— *New York Sun*, November 30.

I have to warm to Capt. Allen because, in rising from cabin boy to captain, he paralleled the career of my own great-grandfather. I agree with him that the animals must have come from Florida, or rather, the Caribbean zone, but disagree that they belonged to a deep water species. It is is far more likely they were lost - swept out to sea and caught up in the Gulf Stream. On the other side of the world, salt water crocodiles have been known to end up as far afield as Fiji. Although no limbs were mentioned - and it is hard to see how they could have been overlooked at such a short distance - the animals were clearly crocodilians (crocodiles, alligators, or caimans), of which the Caribbean coasts shelter several species. But which species?

Judging size and distance at sea is tricky at best, and fear or astonishment tends to lead to over-estimations. Nevertheless, at the extreme close range in this case, one would tend to trust the witness's estimation. But even if we trim off 20% of the length from 7½ to 6 metres, we are still at a size which even a saltwater croc from southeast Asia or Australia rarely obtains. In the Caribbean, a big American alligator, *Alligator mississippiensis* might exceed 4 metres. Most other crocodilians in the region are no bigger. The Orinoco crocodile, *Crocodylus intermedius*, as its name implies, belongs the Orinoco River system more than the coast, and is now rare, but the large males can reach 4 or 5 metres, and one is reputed to have attained nearly 7 metres.

Thus, we are left to hypothesize a combination of unlikely events: an exceptionally outsized male of a riverine species swept a couple of thousand miles out into the ocean - and accompanied by a juvenile. Female crocodilians (not males) care for their hatchlings, but the young are on their own long before they reach this intermediate size. Also, one wonders why the current failed to separate them.

Chapter 2

The 1870s and '80s

West Africa, 1871

This sighting apparently took place in late October 1871, but was not reported until the captain was back in his home port on another ship. The news reached Australia on Saturday 17 August 1872, where it was published on page 3 of both the *Evening News* (Sydney) and the *Age* (Melbourne).

The Great Sea Serpent Again.

Captain M'Taggart, of the ship Kent, at present in Liverpool, reports that he left Liverpool for Benin, on the West Coast of Africa, in the brigantine Onward, on the 20th September last, and that when about 60 days out, and when he was between Cape Palmas and Grand Bassa, one night the vessel was surrounded with enormous shoals of fish of every description, including sharks, porpoises, &c., and although he had been trading on the coast for upwards of twenty years, he never saw such a sight before. Next morning, about eight o'clock, on going forward to take the sun, he observed something in the water, on the Starboard bow, and he at once called the attention of the crew to it, and they, and the officers of the Onward, at once pronounced it to be a sea serpent. As far as Captain M'Taggart could judge, the head, which was very broad, and surmounted by something shaped like a coronet, was about eight feet [*2½ m*] out of the water, and it was going through the water at a very rapid rate, knocking the sea-spray about like a ship. The strange fish went on rapidly for about two minutes, when it stopped and remained stationary. This gave the captain time to observe the fish more minutely. About ten feet [*3 m*] from the head there was a large fin about two feet [*60 cm*] out of the water, and further on there was another about one foot out of the water. The scales were large, and of a beautiful colour. From the head and shoulders, which were of immense width, the body of the fish tapered gradually away to an extent of about 180 to 200 feet [*55 to 60 m*], ending in a tail something like that of a mackerel. In fact, Captain M'Taggart says the colour of the fish clearly resembled a mackerel. After lying quiet for some time, the fish or serpent shot ahead again at great speed, and was soon lost to view. The captain thinks that the presence of such vast shoals of fish on the night previous so far out to sea must have something to do with the presence of this monster on the African coast.

North Atlantic, 1873

I'm assuming the Western Islands in the article refer to the Outer Hebrides of Scotland. Therefore, the site of the incident must have been a bit west of Ireland, perhaps about 55° N, 15° E, give or take a few degrees. A major problem is that it wasn't reported until 18 years after the event. This comes from the *Tasmanian* (Launceston), Saturday 12 September 1891, on page 4.

The Sea Serpent

Yet another sea serpent story has come to light, this time through the columns of an Adelaide journal. A contributor signing himself W. J. Horawell, relates his experience of this much discussed monster as follows:- In the year 1873 and the month of September, whilst on the return passage from the coast of Chili [*sic*] in the bark Glanrafon, 472 tons, J. Sharp, master; owners, Messrs Richardson and Co., Swansea; and being about 300 miles [*500 km*] W.S.W. of the Western Islands, therefore subject to the influence of the Gulf Stream. The ship was not making more than three knots per hour [*5½ kph*], although the wind was several points free and the yards checked in. During the morning we had seen many large pink objects near the surface and around about us, but none close enough for observation, or to ascertain what they were without going out of our course. About 11 30 a.m. I was sent aloft to arrange the gear for settling the maintopgallant studding sail, and seeing one about quarter mile distance and few points on the starboard bow, I reported it to the master, who ordered the helmsman to luff up but keep the sails full, with the object of passing close to it. Unfortunately this order did not permit of our passing within sixty feet [*18 m*] of it, although some of the sportsmen on board were prepared to harpoon it, with a hope of making a closer acquaintance and solving a mystery. Being still aloft and on the upper topsailyard at the time of passing it, and from this elevated position I was situated most advantageously for observing its form and size and on sketching it afterwards to another apprentice (W. White) who was on the mainyard, said it appeared somewhat different and this difference was accounted for by my being about 25ft [*7½ m*] above him. The general impression of those on deck was that its length was 40ft, but I think this was in excess of what should be seen by them, for they were unable to discern what I presumed to be the tail turned back over the body for about a third of its length. Therefore I am safe in saying its full length was 47ft [*14.3 m*], and at the broadest point 10ft [*3 m*]. The master and all hands regretted we had not passed close to it.

With as little detail as this, at least it is unlikely he was making it up! Personally, I would not regard a width to length ratio of 21% to be very serpentine. This alone would suggest it was some sort of whale. It would not have been the last time a whale had been confused with a sea serpent. But what are we to make of the tail flipped back over its body for a third of its length? That doesn't sound very whale-like. And was it really pink? I haven't heard of any whales, or sea serpents, for that matter, of that colour. And how come there were not one, but "many large pink objects"? It is all very mysterious.

Maine, USA, 1875

This is another one from the coast of New England. It was first reported in Australia on page 27 of the *Evening News* (Sydney) on Wednesday 21 July 1875, so the incident may have take place the previous week.

THE SEA-SERPENT AGAIN.

A sea-serpent story that gets a little ahead of the ordinary run of such yarns comes from Portland. Captain Oliver, of the Path schooner Winslow Morse, relates that Thursday midnight, when off Cape Elizabeth, about 15 miles [*24 km*] south-east, while he was at the wheel and another man stood on deck, a great snake rose out of the water, about four feet [*1.2 m*] above the rail, and the body appeared about as large round as a hogshead. The man on deck picked up a long pole with a pike-head on the end, lying near by, and thrust it into the monster's body. The serpent immediately dived and came up on the other side of the vessel, a little distance off, and glided away, making but a slight ripple on the water. It appeared about 120 feet [*36 m*] long. The captain exhibits the pike covered with the blood and sinews of the monster — or some other animal.

Mid-Atlantic, 1877.

Although the sighting by the crew of the *Sacramento* took place in the mid-Atlantic, it was not reported for another three months, when the ship reached port in Melbourne. This is a sighting which did find a place in Heuvelmans' book, but since I have access to the original report, I might as well cite it. The story seems to have been reported in most of the major Australian newspapers, with varying degrees of accuracy and detail, but the earliest and best appears to have been in *The Age* (Melbourne), Monday 22 October 1877, on page 2.

ANOTHER SEA SERPENT

The great sea snake has again turned up, having been seen quite distinctly by the captain of the ship Sacramento, which arrived on Saturday from New York. The following is the extract concerning the sea monster from Captain Nelson's log-book:- 30 July, in 31.59 north and 37 west. Was called on deck by the man at the wheel, who stated that a great snake was passing a short distance from the vessel. On reaching the deck the monster was plainly visible, moving steadily through the water, propelling itself by two large flappers or fins, situated a short distance behind the head. It was about forty feet [*12.2 m*] long and the girth of a beef barrel, the color being yellow. The man at the wheel states that he distinctly saw the eyes of the animal when he first called the captain, who says that he did not hurry up when he was called, as he did not believe in the existence of such a creature; but when he did get on deck, he saw it clearly enough to be certain it was a living animal.

Interestingly, the *Australasian Sketcher* got in contact with "the man at the wheel", and under his direction, produced a front page sketch of the animal in its issue of Saturday 24 November, 1877. Here is a copy of the front page. For those who can't read the text, it says:

THE SEA SERPENT, AS SIGHTED BY THE SHIP SACRAMENTO: FROM A SKETCH BY THE MAN AT THE WHEEL.

"This is a correct sketch of the serpent seen by me while on board the ship Sacramento, on her passage from New York to Melbourne,

I being at the wheel at the time. It had a body of a very large snake; its length appeared to me to be about 50ft. or 60ft. *[15¼ - 18.3 m]*. Its head was like an alligator's, with a pair of flippers about 10 ft. *[3 m]* from its head. The colour was of a reddish brown. At the time seen it was lying perfectly still, with its head raised about 3 ft. *[90 cm]* above the surface of the sea, and as it got 30 ft. or 40 ft. *[9-12 m]* astern it dropped its head. - JOHN HART."

? Ryukyu Islands, Feb/March 1879.

I have no idea what site was called Rock Island in 1879, but since it was located on the route from Hong Kong to Japan, I suspect it was part of the Ryukyu chain. This report is taken from the *The Express and Telegraph* (Adelaide, SA) of Wed 21 May 1879, on page 2.

The sea serpent is continually turning up. No sooner does the interest excited by the tale of his having been sighted begin to die away, than, like the clown in the circus, "Here we are again." The following is an account of his latest appearance, taken from the *Japan Herald* of March 4, and furnished by a correspondent to that paper :- "While proceeding from Hongkong to your coast on board the steamer Radnorshire, when about eight miles W.S.W. of Rock Island, I saw what was undoubtedly a veritable sea serpent. We were proceeding at from ten to eleven knots an hour [*18 - 20 kph*]. Whilst standing on the quarter-deck I observed on the port beam, about a ship's length off, what appeared to be a long whitish-brown sack floating near the surface, parallel with the ship. It first of all struck me it was a dead man sown up in a hammock, and having my binocular at hand I was not long in examining it. Just as I had got my glasses to bear on the object, it raised its head clear out of the water about six feet [*1.8 m*], evidently examining the vessel, and I at once saw it was a sea serpent. I ran forward to get one of the officers to verify the fact, and when I returned it had disappeared. The head and neck were dark, and resembled a swan's neck gracefully arched, with an asp's head tapering to the mouth; the body was of a whitish color, and the diameter must have been at least 12 inches [*30 cm*]; the length I cannot vouch for, further than it seemed to be quite 20 feet [*6 m*]. The animal was heading aft, and on enquiry I found that one of the seamen had seen it an hour previously on the starboard side, which implied that it had been cruising round us from curiosity, whilst it was evident that its propelling power much exceeded ours. I don't believe any of the officers saw the animal a second time but myself; however, an intelligent Japanese did do so, for he was loud in his description of the extraordinary fish after it had disappeared."

You will note how the sighting is related without much drama. The colour scheme seems unusual, but readers will be aware of the many, many cases of a thin neck being raised above the surface. There is no known marine animal capable of doing this.

Fiji, August 1879.

This took place off the small island of Cikobia, located at 15° 44' S, 179° 55' W. It is pronounced "thee-kom-bee-a", because Fijian orthography was created by missionaries for the use of Fijian speakers rather than English speakers. The sound of "the" in "thee" or "there" is represented by *"c"* because why use two letters when one will do? And *"c"* is otherwise unused. Also, because "b" is always pre-nasalised ie it always has an "m" in front of it, the letter *"b"* is used for the combination, "mb". This report is taken from *The Sydney Morning Herald* of Wed 27 August 1879, page 6.

> We have Fiji papers to the 16th instant, from which we extract the following: - Some strange monsters are heard of round our coasts (says the *Times*). In the neighbourhood of Cikobia, we are informed, the natives have seen a fish with long eel-like head, of great length, and with a thickness of body larger than the largest post ever put in a native house. This would seem to point to a possibility of solving the mystery of the sea serpent in these waters. At all events this dark animal must have a wide range of habitat, since it has been seen by the wonder-loving fishers of the wintry coast of Gailway, as by the no less passionately fond- of-marvellous story-telling-natives of these sunny isles. Nor is the sea serpent the only marvel, the cuttle-fish has shown itself therein all its giant strength with head of prodigious size, and arms of enormous length and power, a dangerous follow to meet in sailing some hooker among these reefs. And one more sea monster the shape of a crab, of the size of a turtle, from whose terrible claws the spectator fled in dire fear. Naturalists had better send a Commission to Vanua Levu.

That doesn't appear to be much in the way of details.

Near New York, September 1879.

The following comes from the *Evening News* (Sydney), Thursday 30 October 1879, on page 3.

That Sea Serpent

He has been seen again, and this time so near New York as to cause serious alarm among the shore dwellers on Long Island and New Jersey. His presence on the bounding blue ocean was regarded with only a curious interest, but now that he is "in shore" the matter becomes serious. When he gets down to latitude 30deg 35min, longitude 71deg 21min, the thing begins to look serious, and it would not be astonishing in a few days to hear from the gallant crew of one of New York's garbage scows of his presence in the lower bay. There can be no doubt of his having been seen this time, for the official log of the barquentine Falmouth, arrived here from Dundalk, Ireland, on the 3rd instant, contains mention of him. The mate, who saw him, is a man of steady habits, and is "never, never drunk at sea." The captain is also a sober man and he saw the serpent, and so is the cook, who also saw the monster. He is stigmatised in the log as "a snake", and this fact induced the reporter, who was sent to glean some information of the horrid thing, to ask the mate if he might not have been in his boots instead of the sea. The mate, who is a matter-of-fact seaman, whose mental habits tolerate no metaphors or hyperboles, when asked if the snake might not have been in his boots, declared that such a thing was impossible, for, at the time the "snake" was seen he was in his bare feet. The log of the Falmouth narrates that about 4 o'clock on the afternoon of September 2, in longitude 39deg 35min, the ship passed within 20 feet [*6 m*] of a snake, which was seen about 5ft or 6ft [*90 or 120 cm*] below the surface of the water. It appeared to be about 20ft long, had a square-shaped head, and was covered with diamond-shaped scales. At its greatest diameter — just behind the head — it seemed to be about 6in [*15 cm*]. Remarking concerning the other reported sights of a sea monster, mate Cutting remarks that the mariners who report having seen 90ft and 100ft [*27½ and 30½ metres*] snakes must have had bad eyes. He says he has frequently seen snakes at the mouth of the Mississippi, but he never saw a snake in his life as long as this one. It is the first snake, too, he says, that he ever encountered in the open sea. He is inclined to think that the thing he saw is the identical serpent which has of late been reported so frequently. The snake did not seem to concern himself about the Falmouth at all, but continued to glide ahead as though nothing had happened. Its progress was very slow, and in a few minutes it was far astern.

You will note that two figures are cited for the latitude. The second, 39 degrees, is clearly the correct one, because it is close to New York, which the ship reached the following day. The New England sea serpent was famous during this period, but it typically presented as an elongated animal with a string of humps. On the other hand, this one was clearly a genuine snake. It sounds suspiciously like a python which had been swept well out to see. However, the problem is, it was too far out. Also, an air-breathing land reptile would hardly be swimming 5 or 6 feet below the surface. (It must be pointed out, of course, that real sea snakes occur only in the Indian and Pacific Oceans, and are nowhere near as big.)

Prince Edward Island, Canada, 1879.

This story sounds a bit dramatic, to say the least, and if it is genuine, then it is likely the length of the animal was grossly exaggerated due to the witnesses' fear. It allegedly took place near a small village on the western tip of Prince Edward Island. Although it was said to have occurred in August 1879, it was not for another five months before the Australian press picked it up, after which it did the circuit of various country newspapers (not big city dailies) for another four months. The earliest was on page 4 of *The Ballarat Star* (Vic) of Sat 17 January 1880.

A FISHING BOAT ATTACKED BY A SEA SERPENT.

The *Sumerside Journal* (Newfoundland) publishes the following, received on the 23th. August, from a correspondent at Miminigash:—"At Miminigash, on the 16th of August, as Matthew M'Donald and James Doyle, two men in the employ of E. G. Fuller, were hauling their trawls, they observed an unusual commotion in the water near them. 'Is that a squall?' said Doyle. 'Great heavens!' exclaimed M'Donald, as the line he was hauling took a surge and parted a hook, tearing his hand from one side to the other, and a huge form arose from the sea full 20 feet [6 m] out of the water. 'Quick with the sails, Jim,' cried M'Donald, and the two-terrified men spread their canvass in a hurry. M'Donald gave the helm to Doyle, who, crouching down in the after-berth, barely showed his head, while Mac, rather the cooler of the two, quickly improvised a sort of spear out of a long knife, which he lashed to an oar. He described the fish as a sort of a snake, striped yellow and white, with a mouth as large as the opening of a puncheon, and each time it raised out of the water it uttered a sort of roar like the bellowing of a bull. As the boat, with twice as much sail as was consistent with safety, was flying before the stiff S.W. breeze, the monster followed in her wake. M'Donald thought to pacify it with fish, as it was doubtless enraged by being torn by the trawl hooks, so he commenced throwing hake, with which the boat was partly loaded, to the monster, who greedily devoured them. Nearing the shore they crossed a lobster trawl of Mr Belyea, and the monster fouled and parted it, half filling the boat at work on it with water. This seemed to infuriate it, and raising itself in the air, it made a rush for the boat. M'Donald says he thought all was up with him, but he kept cool, and raising his improvised harpoon, struck the monster in the eye, driving the oar clean into its head, and breaking the knife in the wound. With a roar of pain it sank out of sight, reddening the water around with its blood. Doyle says he counted twelve sharp fins on it, each surmounted with a sort of horn, and both men say the fish was 200 feet [*60 m*] long. To-day, while repairing their broken line, they took off one of the hooks a large tuft of yellowish hair attached to a piece of skin resembling pigskin, which may be seen at Mr Fuller's establishment.

Newfoundland, 1880.

The following is taken from *The Capricornian* (Rockhampton, Qld) for Saturday 31 January 1880, on page 15. Nevertheless, I cannot be certain that the event actually took place in that year, because there was frequently a significant lapse before the Australian press picked up the reports. *The Capricornian* appears to have been the only newspaper to carry the report, and it attributes it to an "American Paper".

THE MONARCH OF THE SEA.— When off the Newfoundland banks recently, an officer of the steamship Anchoria noticed a disturbance in the water about a mile [*1.6 km*] distant on the port beam. At first he thought the commotion was caused by a school of porpoises, but, on closer observation he changed his mind. When he looked through a pair of strong glasses be saw the head and body of the sea serpent rising above the water. Portions of the back of the creature could be seen rising out of the sea at intervals as it propelled itself along on the top of the water. Its motions were similar to those of the land snake as it moves along on the ground. The water in the wake of the creature had been lashed into foam by its tail. Its head was large and contained an enormous mouth, which opened frequently and spat out large quantities of water. Its tongue, which was extremely long, could be seen at times, but no teeth or fangs were observed. The body of the serpent was round, and its colour was black. It was moving in the same direction as the steamship, and at a greater rate of speed. When the creature had got a little ahead of the vessel it sank down into the water and disappeared. The officer estimates it to have been fully as long as the steamship, which is 408 feet [*124 metres*] in length. — *American Paper.*

The length has clearly been exaggerated - or else, the wake was confused with part of the tail. The enormous head spitting out water suggests to me that it may have been a baleen whale, despite the reference to an extremely long tongue.

Orkneys, 1885.

This brief report was apparently taken from the *Scotsman*, and published in the *Weekly Times* (Melbourne) of Saturday 5 September 1885, on page 4.

The Sea Serpent Again.

The sea serpent has been seen in Orkney waters. A number of men well known in Kirkwall, and whose veracity is above suspicion, upon going over to the point at Scargum, near Kirkwall, to fish, saw moving in the water a huge monster, which they described as follows: It was about 100ft. [*30 m*] long, round in shape, and covered with brown hair. When seen it was swimming with great rapidity, and those who observed it were so frightened that they ran away. The report has caused great excitement in the town; but a local naturalist gives it as his opinion that the supposed monster must have been a shoal of seals following each other in Indian file fashion, which, he says, is not an uncommon sight during the breeding season. — 'Scotsman,'

I suspect the "local naturalist" might have been right, because very few sea serpents are described as being covered with brown hair. However, a shoal of seals would give the impression of a row of humps rising and falling. It was a pity the witnesses weren't asked such obvious questions as how the "monster" swam, what the head looked like, the distance, and the duration of the sighting - all of which appear to be of little relevance to certain journalists.

Massachussetts, 1886.

At this period the "New England sea serpent" was gaining a certain notoriety. Heuvelmans had apparently heard a brief, second hand account of this case, because on page 290 of his book occurs the following footnote:

> Some witnesses were respectable: G. P. Putnam, principal of a Boston grammar school and a namesake of the well-known publisher, saw one off Gloucester . . .

Here, then, is the full account, taken from page 2 of *The Express and Telegraph* (Adelaide, SA) of Tuesday 12 October 1886, which is the earliest record I could find.

A Fine View of the Sea Serpent
A Boston Schoolmaster Gives a Detailed Account of the Animal.
The Serpent Seen by 50 People.

Renewed evidence has been given during the past week (says a recent issue of the Philadelphia *Weekly Press*) by trustworthy persons that something like a sea serpent is off the Massachusetts coast. Mr. G. B. Putnam, master of the Franklin School, Boston, writes from Pigeon Cove:—This afternoon about 1.15 o'clock a lad, the son of Calvin W. Poole, was seated upon the rocks near the "Linwood" when something in the water attracted his attention. He immediately ran to his father, who was surveying near by, and, pointing to the object, called out—"The sea serpent! The sea serpent!" Mr. Poole brought his powerful glass to bear upon it, and was at once satisfied that it was the veritable sea serpent. He was about one-fourth of a mile [400 m] from the shore and about two miles [3.2 km] from where he was seen last month. He was moving slowly in a northerly direction. It was a dead calm, a smooth sea, with a bright sun shining, so that there was the best possible opportunity to observe his motions. The distance passed over while visible was at least a mile [1.6 km], and the time occupied not far from twenty minutes. Mr. Poole called my attention to his snakeship at once, and as he passed directly by my cottage I was able with an excellent marine glass to observe his movements, which corresponded exactly with those described by Dr. Sanborn last month, as well as those given in *Harper's Monthly* some years since, also the accounts given of one seen in Gloucester Harbor about 1817. The head was frequently raised partially out of the water, and the movement was a vertical one, showing some 10 or 15 ridges at once. I should say that he was at least 80 feet [24 m] in length. There were perhaps 50 persons who witnessed the passage in part at least, among whom were Samuel Bullock, master of the Bunker Hill school, Boston; Professor Stephen Emery, of the N.E. Conservatory, with his family; Capt. Jackson, the artist, James Hurd, and several guests at Linwood, as well as four members of my own family. I shall never doubt that the sea serpent is a fact.

Connecticut, March 1888

The nineteenth century saw a lot of sea serpent activity around New England. In this case, it was close to the Cornfield Point Lightship, which was positioned at the mouth of the Connecticut River from 1856 to 1957. The earliest account I discovered was in the *Maitland Mercury and Hunter River General Advertiser* (NSW) of Thurs 7 June 1888, and it cited an undated edition of the *N. Y. Herald*. However, I have decided to take a later version which is almost identical, except that it provides the date. It is the *Victorian Express* (Geraldton,W.A.) of Saturday 16 June 1888, on page 3. You will note that these are not exactly capital city, large circulation journals.

AGAIN THE SEA SERPENT
ONE HUNDRED FEET LONG AND AS ROUND AS A FLOUR BARREL

Stonington, Con., March 29. - The schooner Coral, Captain Sherman, is now at this port awaiting favorable weather to proceed to Greenport. To your correspondent Captain Sherman reported that on Friday last [*23 March*], when his vessel was in the vicinity of Cornfield Lightship, there suddenly appeared astern and not two hundred feet [*60 m*] away an immense sea monster that fully answered the description previously given of sea serpents. Captain Sherman say he had a perfect view of the monster. He described it as being over one hundred feet [*30 m*] in length, and in some portions its body was as large round as a flour barrel. The head of the serpent resembled that of an alligator. The captain called his mate, and they both watched the animal until it passed out of sight in the direction of the mouth of the Connecticut River. It passed over the water at quite a rapid gait, and as almost the entire body was on the surface of the water the men had a good view of the creature, and both feel confident that they saw a veritable sea serpent. Captain Sherman appears to be a thoroughly reliable man, and has been master of a vessel for thirteen years, during which time he has made several voyages to the Grand Banks, where almost all species of animals that inhabit the sea are to be found, but never before has he seen anything like the monster above referred to.

South Atlantic, August 1888

This account comes from the *Age* (Melbourne), Saturday 15 December 1888, page 11.

THE SEA SERPENT SURPASSED

The sea serpent is completely thrown into the shade by a creature which has been seen in latitude 39.50 south and longitude 00.13 west. Mr. Webster, chief officer of the ship Bienvenu, has written to the Calcutta papers saying that on August 4, when the ship was hove-to in a strong gale, the second officer, Mr. Parsons, who was on watch, reported that between 4 and 5 in the morning a tremendous living monster passed the port side of the vessel measuring nearly as long as the ship itself. It had large wings or ears, two immense humps on its back 14 or 15 feet [$4½$ m] high, and a tail like a whale. The monster had two eyes the size of teacups, while its back was covered with shells or scales resembling barnacles as large as cocoanuts. It remained near the ship for 20 minutes, its movements being very slow. It is to be hoped that a party of scientific men will, without loss of time, charter a ship and proceed in pursuit of the monster. It differs so entirely from all other described creatures that its capture will create a prodigious sensation, for even the extinct prediluvian monsters were shapely and agreeable to look upon in comparison with this uncouth beast, with large wings or ears, and two immense humps. The public will be glad to learn how many of the Bienvenu also saw this creature, and whether their impressions correspond with those of the second officer. In matters of natural history the minutest accuracy is important. - *Standard*

I am sure we can all agree on the last sentence. It is not clear whether anybody but Mr. Parsons saw the monster but, considering it was in view for 20 minutes, it is likely there were others. Nevertheless, the wee small hours of the morning are not a good time for visibility, and the strong gale would imply a choppy sea. I suspect this is an exaggerated description of a humpback whale, *Megaptera novaeangliae*. I've noticed the same mistake in Australian waters. The "large wings or ears" were the very long pectoral fins which give the whale its scientific name. (*Megaptera* means "big fin".) Its body also possesses large numbers of pale rugosities, "as large as cocoanuts", which serve as anchorages for barnacles. And, of course, it does have a tail like a whale. The incongruities are its being almost as long as the ship, and the two very tall humps. However, if the tail was seen, we must assume it was swimming, albeit slowly, by flexing its body, and who knows how that would appear under the lighting conditions?

Georgetown, South Carolina, 1888

Did a sea serpent turn up in Georgetown Harbour in 1888? On page 573 of *The Great Sea-Serpent* (1892), A. C. Oudemans features a list of hoaxes, culminating at the bottom of the page with:

> The sea-serpent is distinctly seen in Georgetown Harbour, on the 20th. of August, 1888, sleeping on the surface, &c. - *Chambers' Journal*, 1888, Nov. 24. - (R. P. G.)

The three last initials stand for Mr. R. P. Greg, who provided him with his whole collection of clippings. But why was it classified as a hoax? Did he know something he wasn't telling?

Needless to say, Bernard Heuvelmans followed it up, but also thought little of it. On page 227 of *In the Wake of the Sea-Serpents* (1958), he writes simply:

> There was also the sea-serpent which was seen sleeping peacefully in Georgetown Harbour in British Guiana on 20 August 1888;

to which he adds the footnote:

> The date is as doubtful as the sea-serpent, since I have not been able to find the original report. The issue of *Chambers' Journal* cited by Dr Oudemans mentions Georgetown but no sea-serpent.

It doesn't look good, does it? However, in *Mysterious Creatures: A Guide to Cryptozoology* (2002), George M. Eberhart states:

> On September 25, 1888, Captain Springs of the tug *Henry Buck* was towing a schooner in Winyah Bay, near Georgetown, South Carolina, when he spotted a 50 foot animal swimming on the surface with its head 3 feet in the air. The head was vermilion, and

the neck was covered with a long mane. The captain's story was corroborated by others.

For this, he cites "The Sea Serpent", *St. Louis Globe-Democrat*, September 27, 1888, p. 6. This sounds like a reference much closer to the actual event.

Never fear! The original reports have serendipitously fallen into my hands - albeit unmoored from their original source. You will have seen this happen before. A story is published on the other side of the world, picked up by newspapers in the general locality, and then, by a process known only to journalists of the era, vanishes into the ether, whence it is again picked up, months after the event, by some obscure rural Australian newspaper, which was presumably more up-to-date with events in its own vicinity.

This story therefore comes from the *Western Star and Roma Advertiser* of Toowoomba, Queensland, on page 4 of the issue of Saturday 1st December 1888.

The Sea Serpent
A MONSTER OF THE DEEP IN FRESH WATER.
CAPTAIN A. A. Spring, of the steam tug H. L. Buck, brings a strange story from Georgetown. On Saturday his tug had in tow the schooner Jessie Rosaline, on her way to the bar, and had just passed the wreck of the Harvest Moon, which lies in the edge of the channel, when a little boy, seven or eight years old, son of Mr. C. W. Forster, directed his attention to something in the water over the port bow, and asked if it was a bird. Being in charge of the wheel, he paid little attention to the child's question, merely glancing in the direction indicated. He noticed what at a glance seemed to be some large bird floating in the water. He heard the mate of the tug whose attention had evidently also been attracted, remark that it looked like the back of a drowned black. When passing abreast of the object his attention was again called to it. The boat was moving rapidly through the water, so that when he had secured his glasses the object was about two hundred yards away. He examined it intently and carefully, and made out nearly its entire shape. It seemed resting or sleeping, the head and body being more or less exposed to view as the waves rose and fell about it. The mouth appeared to be beak-shaped, the head oval and quite large. The body looked to be as large as a flour-barrel, and

lay upon and in the water in the curves common to snakes while swimming. The tail was not at first entirely visible. While looking intently at the monster something - possibly the noise of the tug - seemed to arouse it, and in an instant it threw its tail into the air, exposing fully fifteen feet [*4½ metres*] of its length, and lashed the water into foam. It swam off in the direction of what is known as Muddy Bay and the Mud Flats, where it was impossible for the tug to follow it. The colour of the monster was very dark. As well as could be judged, the portion of his tail lifted from the water was eight or ten inches [*20 to 25 cm*] in diameter, and his estimated length thirty feet [*9 m*]. The captain of the schooner, who got a much nearer view, estimated the monster's length at fifty feet [*15 m*]. As the point where it was seen the water is fresh, as it is several miles below, and Captain Spring thinks the animal was made sick by it, and if he does not find his way back to salt water very soon his life will be the forfeit for his rash visit to port, and science may yet have an opportunity of fixing his identity.

You will note that the article has been taken, without any alternation, from some publication for which the context was obvious, but which provides no indication as to the precise date it occurred, or at which Georgetown. For all the reader might have know, it could have happened the previous Saturday at George Town, Tasmania!

More than two months later - on Saturday 9 February 1889 - the same story, but told by a third party, appeared on page 4 of an even more obscure Australian newspaper, the *Kadina and Wallaroo Times* of South Australia. This time it was clear that the event took place in South Carolina, because of the proximity of Charleston. This time, too, it tells its source: some mysterious publication called *Iron*.

A NEAR VIEW OF THE SEA SERPENT.

We wonder how many more times the sea serpent will have to be killed before it is finally exterminated. The latest apparition of the hydraheaded monster has this time been in American waters, where it made its *debut* on Thursday, August 20. Captain Hubbard, of the steamer Planter, plying between Charlestown and Georgetown, reports that the sea serpent was seen in Georgetown harbour on that day, half-way between the port and the bar. This is what he says:- "The tug Henry Buck passed within two hundred yards of the monster, and the captain examined it carefully with his glass. He says he made out nearly its entire shape. It seemed to be resting or sleeping, the head and body being more or less exposed to view as the waves rose and fell about it. The mouth appeared to be beak-shaped, the head oval and large. The body looked to be as thick as a flour barrel, and lay upon in the water in the curves common to snakes while swimming. The tail was not visible. While looking intently at the monster, something (possibly the noise of the tug) seemed to arouse it, and in an instant it threw its tail into the air, exposing fully 15ft of its length, and lashed the water into foam. It swam off in the direction of what is known as Muddy Bay and the mud flats, where it was impossible for the tug to follow. The colour of the monster was very dark. The length is stated to be about 50ft. That portion of the tail lifted above the water was between 8in and 10in in diameter. At the point where it was seen the water is fresh for several miles below, and Captain Springs, of the tug, thinks the animal was made sick by it. It is thought that the monster cannot get out of the harbour. As soon as the news was received an expedition as made up to go in search of it, and it is possible that the sea serpent problem may yet be definitely settled." Up to the time of going to press we have received no solution of the problem; so we suppose the expedition is still in search of the "varmint." - *Iron*

You will note there is no mention of the mane, or the vermilion head 3 feet in the air, as reported by Eberhart. By and large, the animal does not appear to have been particularly outrageous as sea serpents go, and I suspect that the label of "hoax" was premature.

You will also note some discrepancy as to date. The first article says Saturday. The second says Thursday 20 August, but 20 August fell on a Monday in 1888. (It fell on a Saturday in 1887, if that is relevant.) And, in case you are interested, 25 September 1888 was a Tuesday.

I put all this on my blog, and ended up by saying: "Perhaps when someone finds the original source, we will be able to work it [the correct date] out. I then received a message from a Mr. Theo Paijmans, who appears to have performed tremendous work in looking up American newspaper archives, to none of which I have access.

It appears there is not one primary, single source for the story. The St. Louis Globe-Democrat, September 27, 1888 edition was one of many American newspapers that carried the story in the days of September 27 to 29. Here's a partial list from my sea serpent sightings database of other American newspapers that ran the account:

27 September
'Corralling The Sea Serpent', Chicago Tribune, Chicago, Illinois, Thursday, 27 September 1888.
'The Sea Serpent Described', The Sun, New York, New York, Thursday, 27 September 1888.
'The Sea Serpent', The Cincinnati Enquirer, Cincinnati, Ohio, Thursday 27 September 1888.
'That Sea serpent Again', The Atlanta Constitution, Atlanta, Georgia, Thursday, 27 September 1888.
'The Sea Serpent Seen', The Times-Democrat, New Orleans, Louisiana, Thursday, 27 September 1888.

28 September
'The Sea Serpent Described', Evening Star, Washington, District of Columbia, Friday, 28 September 1888.
'Our Old Marine Friend Again', Hartford Courant, Hartford, Connecticut, Friday, 28 September 1888.
'Still Living', Lincoln Evening Call, Lincoln, Nebraska, Friday, 28 September 1888.
'The sea Serpent', The Leavenworth Standard, Leavenworth, Kansas, Friday, 28 September 1888.
'The Sea Serpent's Trip South', The Wilmington Messenger, Wilmington, North Carolina, Friday, 28 September 1888.
(brief mention) Fort Scott Evening Globe, Fort Scott, Kansas, Friday, 28 September 1888.
(brief mention) The Decatur Herald, Decatur, Illinois, Friday, 28 September 1888.

29 September
'The Sea Serpent Seen', The Weekly Times Democrat, New Orleans, Louisiana, Saturday, 29 September 1888.
'A Monster sea Serpent', Waco Morning Star, Waco, Texas, Saturday 29 September 1888.
(brief mention) The Herald-Despatch, Decatur, Illinois, Saturday, 29 September 1888.

'The Sea Serpent's Trip South', New York, New York, The New York Herald, 27 September 1888.
'Letter From Charleston', Baltimore Sun, Baltimore, Maryland, Friday, 28 September 1888.

October
North Carolina, Durham, The Durham Recorder, 3 October 1888.
Pennsylvania, Philadelphia, The Times, Tuesday, 2 October 1888.
'The Sea Serpent's Fame', South Carolina, Georgetown, The Georgetown Enquirer, October 17, 1888.

December
'A Near View Of The Sea Serpent', Milton Bruce Herald, Milton, Otago, New Zealand, Tuesday, 18 December 1888.

The Iron, 'the mysterious publication' that you mention, is the England, London newspaper named 'Iron'. It published the sea serpent account under the header 'Occasional Notes - A Near View of the Sea Serpent' in its Friday, 26 October, 1888 edition.

He also advised that the first appearance of Captain Hubbard's story was on 27 September 1888. Thank you, Mr. Paijmans! It thus appears that the story appeared in multiple publications, but had been left out of the loop for researchers because somebody decided it must have been a hoax, and preferred not to discuss it further.

Rhode Island, 1888
This event took place near Point Judith, which is roughly halfway along the coast of Rhode Island, at 41° 36' N, 71° 48' W. There is a reference to an earlier one at Watch Hill, which is at the westernmost point of Rhode Island. Presumably this was described in some earlier American newspaper. The following story is taken from page 5 of the *Weekly Times* (Melbourne, Vic) of Saturday 10 November 1888.

The Sea Serpent

The sea serpent (says the *New York Sun*) has again been sighted, and the newspaper wits will revive the old, old jests at its expense. After having been discovered off Watch Hill a few days before, the mysterious monster was encountered recently to the south-east of Point Judith by Captain Delory, of the sloop Mary Lane.

But we can tell these sharp young men of the newspapers that existence of the sea serpent is no jesting matter. It has been the subject of serious scientific discussion for many years past, and zoologists are now by no means disposed to scoff at the possibility that there is in the sea a creature such as Captain Delory reports having seen, and which may be a modified type of monsters of past geological periods of which we have the fossil remains.

Captain Delory describes the Point Judith sea serpent as having a head like an alligator's, with jaws that 'looked to be at least five feet [*150 cm*] in length, and were studded with teeth six inches [*15 cm*] long, while the eyes were as large as the crown of a hat. Back from the head ran a huge fin, which was kept straight.' The entire length he estimates as about seventy feet [*21 m*]. This recalls the graphic description of the fossil ichthyosaurus and plesiosaurus in Hawkin's 'Extinct Monsters of the Ancient Earth.'

The rest of the article continues with a discussion of said ichthyosaurs and plesiosaurs. The journalist would have been better off taking his tongue out of his cheek and asking a few questions on such details as distance, time, colour, width, movement, and so forth.

Pacific Coast off Oregon, ? January 1889 or late 1888

This stretch of coast has long been notorious for sightings of an elongated, vertically flexing creature whimsically labeled "Cadborosaurus", and the next report is a perfect early example. It comes from the *Brisbane Courier* (Brisbane), Friday 18 January 1889, page 6.

THE SEA SERPENT AGAIN

The regular annual sea serpent has made his appearance again. He is a little out of his latitude this time, having been seen in a place where heretofore he has never been known to roam. There is no doubt as to the identity of the creature, as it is vouched for by several parties who are known as strictly temperate men, whose eyes have not been accustomed to seeing every variety of snakes floating in the air and in every conceivable position. Captain Edgar Avery, of the barque Estrella, while coming from Tacoma to this city [San Francisco] with coal, descried the monster when the barque was passing the Umpqua River. The serpent, for such the captain solemnly declares it to be, was swimming on the surface of the water in a southerly direction. The barque at the time was headed south-south-east, and when the captain first noticed the reptile it was about 200 yards off, and was apparently not the least disconcerted by the proximity of the vessel. As it was 10 o'clock in the morning, and the sun was shining brightly, the startled captain had a good view of the serpent. When he was satisfied that he beheld a real live serpent, and not a creation of his imagination, the captain sprang below and got his rifle, calling to his wife and crew to come on deck and view the wonder. The lady and several of the crew came on deck and plainly saw the monster swimming by. He appeared to be about 80ft. [*24 m*] long, and as big round as a barrel. He rode over the waves with is head and about 10ft. [*3 m*] of his body elevated above water, every now and then dipping his immense head into the water, the body making gigantic convolutions while gliding caterpillar-like over the waves. The head was flat, or "dished," as the captain described it, and the body appeared to be covered with scales. About 10ft. of what might properly be called the neck was covered with coarse hair, resembling a mane. After viewing the monster for a time, the captain raised his rifle and fired several shots at it, but the bullets fell short. The sea serpent seemingly paid no attention to the shooting, but kept on his way. The excited spectators kept it in view for fully a half-hour, when, without any apparent flurry, it sank out of sight in the sea, and was not seen after. - *San Francisco Alla*

Galápagos Islands, 1889
This is a short paragraph from the *Evening News* (Sydney), Saturday 21 September 1889, page 5.

Here He is Again!
THE SEA SERPENT KILLED.

A dispatch from Panama to the San Francisco CHRONICLE, dated August 1, says: Captain William F. Smith, of the bark Nautilus, reports that when off Cape Berkeley, Galapagos Islands, a sea serpent was seen about thirty yards from the vessel. Captain Smith estimated the serpent's length at 80ft [*24 m*], and he was about as large round as a barrel in the thickest part.

The head was shaped like a snake's, only on the extreme end of the upper jaw there was a ridge or bunch. The head was about 3ft [*90 cm*] in length, and about 2ft [*60 cm*] back of the head was a mane of hair. No fins were seen. The tail was long and tapering, and shaped like that of an eel. The captain and mate loaded two bomb-guns, and banged away at him, and for about fifteen minutes there was quite a circus, the serpent lashing the water with his tail, and running his head out 4ft or 5ft [*1.2 - 1.5 m*]. At last he ran out his head, whisked around, and sank, dead.

Did it really happen? In this book I have deliberately omitted outrageous stories which are clearly hoaxes, including tales about the sea serpent being killed and the body brought back for scientific investigation - stories whose expected follow-ups never occur. I suspect this belongs in that category. The reason, however, that I am not prepared to completely rule it out is that there exists a much better documented case of H.M.S. *Hilary* using a sea serpent for target practice in 1917. I shall spare you my opinion about this sort of behaviour.

Chapter 3
Strange Visitors to New Zealand

Australia may be on the other side of the world to the rest of the Anglosphere, but there is one section of it which is much closer to home, and with which it is in constant communication: its sister colony on the other side of the Tasman. The eastern capitals of Australia were, and are, closer to New Zealand than Perth. We are taught how, in 1901, the six Australian colonies federated to produce a new Commonwealth, but we have forgotten that, prior to 1901, New Zealand was classed as the seventh Australian colony. We should not be surprised, therefore, that the Australian press featured a whole array of New Zealand sea serpents which the rest of the world failed to notice. Let us now examine them.

Tasman Sea, 1877.

This report appeared on page 4 of the *Ballarat Courier* (Vic) of Tuesday 11 December, 1877, and page 2 of the *Avoca Mail* of the same date. It should be noted, however, that the witness appears to have got the dates wrong. According to shipping reports, the *Arawata* sailed from Melbourne on 8th November, not the 7th. Also, the 15th was a Thursday, not a Wednesday. In fact, the 15th was a Wednesday on no other month except August. In any case, the sighting would have taken place on either 14 or 15 November, when it must have been pretty close to New Zealand.

THE SEA SERPENT AGAIN

It appears that this monster of the deep is determined to thoroughly establish among men a belief in its existence, for accounts are continually coming in from persons to whom it has shown itself. Mr Howell, a Sandhurst gentleman, who left Melbourne for New Zealand by the Arawata on the 7th ultimo, writes as follows to a friend in the city: - "We had a splendid passage across, but met with no adventure worth recording till Wednesday, the 15th instant when we saw something that very much resembled the celebrated 'sea serpent,' about which there has been so much of late. It was about seven o'clock in the evening when the chief steward drew the attention of myself and two others to the monster, on the windward side of the vessel. Both he and the cook, who also witnessed it, have been to sea the greater portion of their lives, and state they never saw anything like it before. It rose about 4 or 5 feet [*1.2 to 1.5 metres*] out of the water, and enabled us to see about 20 feet [*6 metres*] of its back which was quite black, and covered with long horns or prongs, 2 or 3 feet [*60 or 90 cm*] in length. It rose five or six times. The steward ran down and called the captain, but when the latter came up, the brute did not rise again. The steward, Mr Taylor, gave me permission to use his name, and he is well known in Melbourne and all round the coast.

This sounds like a form of the "long necked" sea serpent, but I don't know what it meant by "long horns or prongs", which appear to have been attached to the body of the animal.

Poor Knights Islands, 1892.
These lie about 50 km northeast of Whangarei. The largest is situated at about 35° 30' S, 174° 45' E. This account comes from the *Daily Telegraph* (Sydney) of Monday 9 May 1892, on page 5

THE SEA SERPENT AGAIN
AUCKLAND, Saturday. - The Captain of the schooner Madora reports having seen a sea serpent off the Poor Knights. The schooner was within 12 ft. [*3.65 m*] of the monster, which the captain estimated had a length of 90 ft. [*27.4 m*]
He made a sketch of it.

That's not much to go on, and I hope the sketch, with a much more detailed description, was published in the New Zealand newspapers. Perhaps one of my trans-Tasman readers can check.

Timaru, 1893.
Timaru is a port halfway down the east coast of the South Island. This is another report cursed by a lack of details. It comes from page 3 of the *Evening News* (Sydney) of Monday 23 November 1893

Another Sea Serpent.
AUCKLAND, Monday. - The officers of a steamer which arrived in port on Saturday, report having seen a huge sea serpent off the coast to the south of Timaru. They state that the monster lifted its head fully 15 feet [*4.6 m*] above the surface of the water.

Other reports stated 14 feet, rather than 15. It is hard to see how they could be so specific. In any case, it was obviously one of the "long necked" variety.

Cook Strait, 1896.

Cook Strait separates the North and South Islands of New Zealand, and the Brothers Lighthouse stands atop a rocky domed island in the middle of the strait at approximately 41° S, 174½° E. This report comes from the *Age* (Melbourne), Thursday 18 June 1896, on page 6, though it was not the only Australian newspaper to carry it (of course!).

> A SCHOOL OF SEA SERPENTS (?)
> AUCKLAND, Wednesday.
>
> Tregurtha, keeper of the Brothers lighthouse, in Cook Strait, reports that on 29th May he saw seven strange marine monsters about three miles [*5 km*] off. At first he thought them whales, but he could not see any spouting, although there was plenty of water splashing. Three animals rose 15 to 20 feet [*4½ to 6 metres*] out of the water, remaining at that position half a minute, and afterwards a fourth appeared. The heads of the monsters were snakelike, and the necks much smaller than the heads, of greyish white color. Another monster like an enormous shark with two fins behind his head was seen by Tregurtha and another man called Butler.

I'm assuming that Mr Tregurtha possessed a high magnitude telescope because, firstly, it is the sort of thing you would expect of a lighthouse keeper and, secondly, because he would not have been able to distinguish anything significant at that distance without one. Just the same, even with a good telescope, three miles is a long way to estimate the height of any moving object - or even to estimate the distance of three miles. That being said, it is hard to think of any known species which would appear as long necks rising from the water.

Stephen's Island, 1896.

This small island is located just north of the South Island at approximately 40° 40' S, 174° E. The account is taken from the *Chronicle* (Adelaide), Saturday 27 June 1896, page 23. Thus, the sighting was not far in time and place from the previous one.

ANOTHER SEA SERPENT.
Auckland, June 23.
Another marine monster, 30 ft. [*9 metres*] long, is reported to have been seen by the officers of the steamer Mahinapua, off Stephen's Island. It was about a mile [*1.6 km*] from the steamer. Its head was snakelike and its color a greyish white.

Other newspapers cited a length of 33 feet, but what is that at a distance of a mile? So what sort of sea creature looks greyish white, with a snakelike head, and appears to be 30, or 33 feet long? Almost anything! I can't see how a report of such brevity can be of any use whatsoever.

Mangawhare, 1896.

1896 appears to have been a good year for sea serpents in New Zealand. Now we head up to the far northwest of the North Island, to a little village called Mangawhare, situated at the mouth of the North Wairoa River, on the west coast, just a few miles north of 36° S latitude. The initial report in the Australian press turned up on page 2 of *The Mercury* (Hobart, Tasmania) of Friday 4 September 1896.

THE SEA SERPENT AGAIN
The United Press Association of New Zealand has published the following telegram:-
"AUCKLAND, August 28.
"A letter, dated Mangawhare, Northern Wairio [Wairoa], August 23, 1896, is published:- 'We, the undersigned, saw to-day on the west coast, at the back of Mangawhare, at about 12 o'clock, a very strange sea monster about half a mile [*800 metres*] outside of the breakers. It appeared to be the shape of an eel, and reared straight out of the water to a height of about 20ft [*6 metres*], and then fell like a tree, sending up spray to a height of about 40ft or 50ft [*12 or 15 m*]. We saw it rise about a dozen times, and the last time it rose it turned backwards, showing a tremendous open mouth as it fell. It was black on the back, and white underneath. It appeared to be about 4ft. or 5ft. [*1.2 or 1.5 metres*] thick. We watched for some time after, but saw no more of it. About a mile [*1.6 km*] further out we saw a large whale spouting at the same time. - We are, etc., JAMES EVANS, CHARLES FLAVELL, THOMAS BRYAN.'"

It is hard to see how "a tremendous open mouth" could be recognized at that distance. Whales have been known to breech in such a fashion, and the black and white coloration is also typical of many whales. However, the apparent slimness of the animal involved, and the absence of flippers, would militate against such an identification.

Now we move into the twentieth century.

Off Banks Peninsula, 1907.
This story originally appeared in the *Canterbury Times* of New Zealand, which dated the event as "Saturday", but since it was not reported in Australia until late March, it is uncertain which Saturday was intended. As for the site, the Banks Peninsula is just south of Christchurch, on the South Island, and Akaroa is situated in a deep inlet in that peninsula. However, I have been unable to locate Goashore Bay, and the Ninety Mile Beach is now on the North Island. However, I am assuming that in 1907 the name was applied to the beach of Canterbury Bight, stretching southwest from the Banks Peninsula. In any case, the first report in Australia was on *The Richmond River Herald and Northern Districts Advertiser* of Friday 29 March 1907, on page 5.

Another Sea Serpent Yarn
A Lincoln correspondent writes as follows:- "On Saturday a remarkable sight was witnessed by a party of Lincoln residents, Messrs. A. Bartram, C. Howell, W. Bartram, R. Bartram, and H. Howell, who were crossing the hill which divides the Ninety-Mile Beach from Goashore Bay, Banks Peninsula. About two miles [*3 km*] out a large, dark object, at first thought to be a whale, was sighted. Presently, what looked like a line of large birds appeared above the water, followed by an enormous snake-like body, fully 100ft. [*30 metres*] long. The bird-like objects were now seen to be a series of large fins or humps along the entire length of the creature's back. The serpent, or whatever it was, was plainly visible for about half an hour, in which time it travelled a distance of about four miles [*6½ km*]. Unfortunately, it was too far distant for detail examination with the naked eye, as it did not approach within a mile. Being observed, however, from a height of about 300ft. [*90 m*], its movements could be plainly seen as it swam rapidly along, with a winding, eel-like motion. At times it raised its head and neck several feet above the water, the head appearing to be about the same thickness as the neck. When last seen it was making in the direction of Akaroa, and was about half a mile

[*800 m*] from land. During the whole half-hour it was plainly visible considering the distance, and remained most of the time on the surface, but sometimes just beneath it. As the sea was very calm, there was no possibility of having mistaken a large whale, a piece of wreckage, or anything else for what was really seen, moreover in swimming about it was frequently going against wind and swell." - "Canterbury Times"

North Cape, 1932.

In *In the Wake of the Sea Serpents*, Dr Bernard Heuvelmans records a sea serpent sighting by a Mr. Bellamy, and then adds a footnote to page 430.

He also mentions, almost equally vaguely, that a little while before 1933 Captain J. Munroe saw, off North Cape, New Zealand, a 150-foot animal as thick as a steamer's funnel.

Well, perhaps Mr. Bellamy had access to a more detailed newspaper article. As for me, I am limited to the *Newcastle and Morning Herald and Miners' Advocate* of Saturday 25 June 1932, page 7.

SEA SERPENT?
Strange Monster off N.Z.
TRAWLER SKIPPER'S STORY
WELLINGTON (N.Z.), Friday.

Captain Munro, master of a trawler, declares that he and two members of his crew saw a sea monster between Three Kings and North Cape on Monday. The monster, said Captain Munro, rose almost perpendicularly from the water until 30 feet [*9 metres*] of its body protruded. There was no sign of fin or tail, and the body appeared to be as round as the ship's funnel. It broke the water 50 yards away three or four times.

Cook Strait, 1939.

The sea between the two islands of New Zealand became the focus of a number of sightings in the middle of this year. The following brief note comes from page 10 of *The West Australian*(Perth) of Friday 9 June 1939.

SEA SERPENT
Strange Object in N.Z. Waters.

WELLINGTON, June 8 - In a letter to the "Evening Post," Mr H. C. Christian of Terawa, Pelorus Sound reports having seen what he took to be a sea serpent in Tasman Bay last Friday. It appeared, he said, to be be eight or nine feet [*2.4 to 2.7 metres*] long and eel-shaped. The tail protruded from the water about three feet [*90 cm*], being split or comb-shaped at the edges. The head had the appearance of a dog's head and had hair on the top. The serpent turned towards the launch when the craft approached. When the launch got within 40ft [*12 metres*] of it the serpent submerged.

I'd love to see the original New Zealand newspaper. The story also ran the same day on page 8 of *The Age* (Melbourne), which added that Mr. Christian had two companions in the boat, and that the length of the animal could not be estimated, but that at least nine feet were above water. (If he saw both head and tail, surely he could have made a more accurate estimate.) The day before (8 June) the *News* of Adelaide recorded on page 11 that the other two witnesses were Mr. Christian's sister-in-law and a native (ie Maori) workman. It confirmed that eight or nine feet were above water, and that it "moved and submerged, later rising again."

Next we have the following story from *The Daily News* (Perth) of Thursday 15 June 1939, on page 26.

SEA SERPENT BOBS UP AGAIN
WELLINGTON, Thurs.

The dog-headed sea serpent, which a Terawa man claimed he saw week ago in Pelorus Sound has been testified to by others. A Dunedin man says he saw the beast some time ago but did not dare to face sceptics. A Wellington fisherman says he and a friend sighted the serpent in Cook Strait and agreed with the description given by Mr. Christian, of Terawa. At first it appeared to be a partly-submerged tree but closer examination showed it was some strange form of marine beast. He said the neck was about 5ft. [*150 cm*] long and the head like that of a dog with hair on top. It disappeared as their boat approached. 'We did not mention seeing it as people are inclined to be sceptical of sea serpent stories,' the fisherman added.

Another brief note comes from page 7 of the *Daily Mercury* (Mackay) of Thursday, 20 July 1939.

SEA SERPENT AGAIN

The sea serpent off Nelson (N.Z.) has been seen again! A launch party reports that it has a small head covered with hair, two tiny ears, and a long neck which was rearing several feet out of the water. Some distance behind the neck, four feet of tail was seen. As the launch approached, the serpent disappeared.

On Monday 31 July the *Advocate* of Burnie, Tasmania, ran a longer article on page 2 about what appears to have been a different sighting, and it also provides some of Mr. Christian's exact words in his original letter.

N.Z. "SEA SERPENT"
Hairy. Head and Tiny Ears
TAIL CURLED OVER BACK NELSON

(N.Z.), Sunday.- Once again a "sea serpent" is reported to have been seen by a launch party off Pepin Island, Cable Bay, but on this occasion it was no fearsome monster, but comparatively small, little more than 8 ft. long and about the same size round as a large conger eel. The creature was seen by Mrs. E. E. Kawharu, of Burville Island, and other members of a launch party which was proceeding past Pepin Island recently.

Mrs. Kawharu said that when first seen the creature was some distance from the launch and about a chain and a half [*20 metres*] from the shore, basking in the sunshine, and was thought to be a piece of driftwood. As the launch drew near a small head with very tiny ears could be seen perched on a long neck about 2 ft. out of the water. The head appeared to be covered with a growth of hair. Curled over its back was a tail about 4 ft. long, fish-tail shaped at the tip.

The launch got within 15 yards of it when the creature dived and disappeared. Mrs. Kawharu said that it was not fearsome in appearance, but she had never seen anything at all like it before.

Early last month, Mr. H. C. Christian, of Te Rewa, Pelorus Sound, reported that while passing Pepin Island in a launch with his sister-in-law and a young native workman, they saw what appeared to be the branch of a tree sticking out of the water. It had a somewhat swanlike shape, and appeared to be 8 ft. or 9 ft. long.

As they got closer to the object they found that it was alive, and it turned its head toward the launch.

"We noticed that its tail, which protruded about 3 ft., curved back toward the centre of the fish, and was split or comb-shaped at the edges," said Mr. Christian. "The head was somewhat like that of an eel, but rather of the appearance of a dog's head, with hair on top, and was 3 ft. out of the water. A most peculiar feature was that when we saw it first it appeared to be quite still, although fully two-thirds of the fish was out of the water, and it remained in that position until I turned my launch toward it, and we got within 40 ft. of it, when it submerged. I have been 35 years in this territory, and have seen nothing like it before,"

The small size initially suggested an eel to me, but the consistent references to tiny ears, hair, and now the neck protruding out of the water would tend to argue against this interpretation.

And now for something light. The following cartoon appeared on page 4 of the *News* (Adelaide) of Saturday 17 June 1939.

FISHERMEN'S TALES

Chapter 4
The Last (Forgotten) Sea Serpents of the 19th Century

Puget Sound, 1891

The following case involves a conventional, honest-to-goodness serpentine shape - in a very conventional setting. Similar sightings off the coasts of British Columbia and Washington have been frequent right up to the present day, so much so that the animal has received the whimsical name of "Caborosaurus", as you will remember from 1889. This report comes from the *Petersburg Times* (South Australia), Friday 11 December 1891, page 3. Note that the story apparently took four months to reach this obscure antipodean newspaper.

A Pacific Sea Serpent

A sea serpent in Puget Sound is the latest sensation. On Sunday, August 2, while rounding Port Williams about 7 o'clock in the evening the Sehome was passed by a huge sea monster from 30 to 40 feet [*9 to 12 m*] long and about a foot [*30 cm*] thick. It was seen by H. B. Street, the boat's quartermaster, and George W. Doney, the pilot. Street was standing near the pilot house when he saw the huge serpent swimming rapidly past the steamer. He did not realize what it was at first, but when it rose to the surface of the water he was rooted to the spot. He says the boat was running about twelve miles an hour [*19 kph*], but the serpent was swimming so rapidly that it passed immediately in front of the bow of the boat and went down on the opposite side.

In describing the scene Street said:

'I first thought it was a seal when I saw its head, but as it rose to the top of the water and I saw about ten feet [*3 m*] of it clear out of the water I knew it was not a seal. Then when I noticed how it lashed the water with its tail I saw that it was a sea serpent thirty or forty feet long, and it left a hundred feet [*30 m*] wake in the water behind it. As it passed around the bow of the boat it lowered its head and spread out a big fin on the upper part of its neck, just back of the head. It swam just like a snake and twisted itself through the water in regular snake fashion. I have been on the water a long time but never saw such a monster before. As soon as I saw what it was I called the pilot's attention to it, and he said at once that it was a sea serpent."

Both Street and Doney are reliable men whose word cannot be questioned, and the fact that they say they saw the monster beyond all doubt establishes the fact there is or was a sea serpent in the water of the sound among the many other wonderful creatures that are found in this arm of the sea.

Florida, 1891

I am unable to provide any original reference to this incident, but I am including it because I have decorated the cover of this book with a contemporary illustration which, nevertheless, bears all the hallmarks of being the work of someone who wasn't there. What is certain is that it is refers to an incident in 1891 at Pablo Beach, Florida, now known as Jacksonville Beach.

On page 194 of *Chronicles of the Strange and Uncanny in Florida* (2010), Greg Jenkins related how people initially felt something slick brush against them in waist-deep water. Suddenly, a scream rent the air, and people started swimming to shore, while onlookers saw something chasing them about twenty yards from shore. It was described as being "twenty to thirty feet [6 to 9 metres] long, about the size of an eel, along with what appeared to be small fins or flippers at its side and black, slimy skin", while its head was round, with a doglike face.

As his source he cited Heuvelmans' book, but this is completely false. All Heuvelmans provided was a copy of the illustration. The original reports must be squirreled away in some local, quite possibly extinct, newspapers of the period. Perhaps some of my American readers with time on their hands can research the matter. You won't need to go through the whole twelve months; I'm sure even in Florida people did not go swimming all year round.

North Atlantic, 1894

I don't like this story. In the first case, the distance appears to have been very great. In the second, as far as I can ascertain, the sighting took place during the morning twilight. But most of all, I don't like its style. It reads more like a short story than a news report. In the course of my investigations, I have come across "sea serpent" stories which do appear to have been intended as fictitious short stories. Nevertheless, this one was presented as if it were intended to be taken seriously as a news report, so I shall leave it to you to make up your own mind. After all, it might just that the journalist decided to insert his tongue into his cheek when he wrote up what the witness said. The source is the *Evening News* (Sydney), Saturday 23 June 1894, page 5.

Another Sea Serpent
STATEMENT ON AFFIDAVIT
MONSTER WATCHED FOR FIVE MINUTES

The sea serpent has been seen again, and by a teetotaller. Discreet and circumspect is First Officer Peters, of the American. The steamer is a "tank". He keeps his tongue from deceit and his lips from grog. So when he said that he had seen a chocolate sea serpent moving over a citron colored sea everyone believed him.

The American left Rotterdam on November 22, bound for New York, with her compartments full of water ballast. The morning of December 2 dawned with lowering clouds and a frost-laden air. A heavy sea was running, and yeasty waves bustled over the scene like the white capped cooks in a hotel kitchen at dinner time.

According to the patent log the tank was in latitude 43deg 55min and longitude 56deg, which is south-west of the Banks of Newfoundland. At 1 min past 7 o'clock Mr. Peters grasped the bridge railing convulsively and craned his neck three points to the port bow.

"Vas ist los mit Peters?" muttered the helmsman.

"Shut up!" roared the first officer. "Look behind you, man, and see that drunken zigzag wake you're making. Lose anything when you came over here the last time?"

"There, sir," said Mr. Peters, "about half a mile [*800 m*] away, three points off the port bow, I saw a great snake more than 100ft [*30 m*] long and as big around as a sugar hogshead. I could just see his back rising and falling in the sea. His head was under water and so was his tail."

"I could see," said Mr. Peters to a reporter, "little of this big fish, and I am sure that it was not a wreck, a whale, a school of porpoises, or a lot of seaweed. I have been at sea, man and boy, these 21 years, and I know that what I saw was a big fish or a snake. He moved with a wavy motion. He bent his back into arches until he looked like a lot of crankshafts. You could see the humps plainly. They rose and fell with a steady beat.

"The steamer was steering west by south and the sea serpent was headed for the south-east. He was of a dark brown color. For five minutes I stood on the bridge watching this great chocolate colored snake wriggling and squirming. When we sighted him he was on the port bow, as I was telling you. When I started down to tell the captain about it the sea serpent was just abeam.

"The captain and I came on deck a minute later and the sea serpent was gone. He had dived and disappeared. We looked for him a long time. All we could make out was the long trail of foam which he had left behind. He seemed to have moved through the water as though he was urged on by a propeller. He left a wake like a naphtha launch.

"What makes me certain that it was a sea serpent? Well, he was not a whale, or he would have come to the surface to blow. He was not a porpoise, for he was not the right color. It could not have been a derelict, for it would not have sunk or left a wake."

Mr. Peters says that one of these days he will capture the sea serpent and discomfit the doubting Mr. Thomases by towing it into port.

Needless to say, if the animal had been swimming in vertical arches, it must have been a mammal. Snakes and eels flex horizontally.

Scotland, 1896

This story comes from an obscure rural newspaper, the *Glen Innes Examiner and General Advertiser* of Friday 21 August 1896, on page 4. Note that, although it dated the event to "Friday night", this could have been any Friday in the previous few months. As for the setting, Redhead is a prominent headland near Arbroath, a town on the east coast of Scotland at approximately $56\frac{1}{2}°$ N, $2\frac{1}{2}°$ E.

NATURALIST
THE SEA SERPENT

The sea serpent has again broken loose, and seems to be disporting himself off the Redhead, near Arbroath, Scotland. The crew of the fishing boat Diligence, which arrived late on Friday night at Arbroath from the deep-sea fishing, reported having encountered a sea serpent off the Redhead on the same evening. One reporter interviewed the crew, and elicited the following narrative of what actually occurred. - the skipper of the boat is Hugh Smith, residing in South Street, Arbroath, and the other members of the crew are Hugh Smith, jun., David Shepherd, Joe. Cargill, and Robert Smith. The skipper states that between five and six o'clock on Friday night, when the Diligence was about seven miles [*11 km*] off land, with the Redhead bearing N.W. by W., Shepherd, the Steersman, called attention to an object which he at first took to be the sale [*sic*] of another boat. The object stood straight out of the water about eight feet [*2½ m*] high, and the head, which was of considerable dimensions, was slowly turned round in the direction of the boat. The skipper, with great boldness, shouted to steer the boat straight for the animal, but this had hardly been done when the monster suddenly disappeared. A minute or two later, however, it again appeared, this time somewhat closer to the boat, and it stood out of the water for a couple of minutes or so, giving the fishermen ample opportunity of seeing it more closely. The body was about 14 inches [*35½ cm*] in diameter and its head was shaped like that of a serpent. What appeared to be the tail was seen at a distance of about 30 yards from the head. The animal appeared to be timid, and no sooner caught sight of the boat than it disappeared. It was, however, seen more than half a dozen times afterwards. As may be imagined, the fishermen were somewhat put about at the strange sight, but they are firmly of the belief that the animal was a veritable sea-serpent.

The fact that the whole crew was interviewed - but probably not separately - suggests this was not simply one man's hoax. The text is difficult to read, but it appeared to state the animal's diameter as *1¼* inches. This, of course, is absurd, and I note that the identical story in a later newspaper gives it as 14 inches, which I have followed. Even so, this precision is not normal for people estimating size at a distance, and I wonder if it were not a misprint. It also sounds surprisingly small for something with a head "of considerable dimensions", and with a tail 30 yards [*90 feet or 27½ metres*] from the head. What a pity the reporter did not ask a lot more questions!

South Atlantic, 1897
The following comes from the *Newcastle Morning Herald and Miners' Advocate* of Monday 6 December 1897, page 8.

THE SEA SERPENT AGAIN

The sea serpent is to the fore again. This time the reptile was discovered by the crew of a cattle ship outward bound from Buenos Aires to Durban, when some of the hands observed what they thought a line of floating seaweed. A few days later the same object was seen a mile astern. The thing remained in sight for several days at varying distances and positions. Then the floating object disappeared for a week. One morning the look-out announced "some wreckage right ahead." The ship in a short time came to "three or four little whales with scaly backs, one behind the other" and "a long, snakey-looking thing with a mouth like a frog - about as long as two right whales, but a good bit thinner." It reminded one of a "monstrous conger eel, with scales like a tarpon." The creature, which was apparently asleep, awoke, and, after rushing through the waves several hundred yards, disappeared. It was seen many times afterwards. The "serpent" invariably dived soon after the carcass of any dead animal was thrown overboard. Two men on a Norwegian timber barque, which reached Durban soon after the steamer had sailed for South America, depose to having seen a similar monster when coming up the coast.

"Two right whales" would imply a length of about 100 feet, or 30 metres. It is not often that a sea serpent has been described as having scales, but it is not unknown. It is also unusual for one to be observed on multiple occasions, or for it to essentially follow a ship. Under such circumstances, it is reasonably certain that the crew hadn't confused a known species with the legendary unknown. Considering the opportunity for lengthy examination, it is a pity that the journalist did not ask for more details.

West of Sumatra, 1899

And now for a real adventure! It was reported on page 2 of the *Launceston Examiner* (Tas.) of Thursday 19 October 1899. Some other newspapers headlined it: "540 Miles in an Open Boat", but the story was the same.

TWELVE DAYS IN AN OPEN BOAT
A SEA SERPENT SEEN - VERY LIKE AN EEL

When the steamship Darius broke her shaft on the voyage to Calcutta, a boat's crew was despatched [*sic*] to Sumatra for assistance. Shortly afterwards the Darius was picked up by the steamer Gulf of Aneud, and towed to Colombo. The following account of the boat's experience (says the "Age") is now at hand:- "By H.M.S. Phoenix, arriving from Batavia, there came to Singapore Mr. Instone (the second officer of the Darius), Messrs. Wilson and Neeson (who were passengers on the vessel), and five of the native crew. When the vessel got into difficulties with her propeller on August 19 it was decided to dispatch one of the boats to make for Sumatra or get assistance from any passing vessel. Accordingly the second officer and a crew of five manned one of the ship's boats, and volunteers being called for, Messrs. Neeson and Wilson also offered to go in the boat. After she was manned and provisioned, she left on the evening of the 19th. A few extracts from the log will show that, although on the whole fine weather and a prosperous voyage were experienced. 12 days elapsed before they were picked up:-

Then follows the boat's log, of which the relevant entries were:

August 30 - At 9.30 p.m. sighted Pulo Bojo light   August 31 - At 6 a.m. sighted a steamer apparently bound south. At 10.30 we sighted a sea serpent about 30 yards from the boat, to all appearances 15ft. [*4.6 m*] in length, and 2ft. [*60 cm*] in girth. The colour of the animal was a dirty yellow, and in shape it was something like an eel. The serpent was right on top of the water, and swan along slowly across out stern.

As an indication of where this took place, you will remember that they sighted the Pulau Bojo light the previous day. At 1 a.m. the following day they passed it. This lighthouse is approximately 200 km west north west of Padang, Sumatra, marking the passage between the Batu Islands and Siberut. It is situated on a small island just off the south tip of another island with the ominous name of Pulau Tanahbala, or "disaster land island".

This is another of those very strange stories. The extreme proximity of the animal, in broad daylight, means the description, presumably written down immediately afterwards, must be pretty accurate. But what was it? The length doesn't sound very sea serpentish. It sounds more like some sort of large dolphin. There are a couple of species which reach that length, but none are anything like dirty yellow in colour, and all possess the sort of fin which observers would hardly fail to mention. We can just put this down as further evidence that the sea still holds rare species unknown to science.

Northumberland, England, 1900.

This brief story comes from the *Maryborough Chronicle, Wide Bay and Burnett Advertiser* (Queensland), Monday 17 September 1900, on page 3.

> The "sea serpent" has again made its appearance, this time off the coast of Northumberland. The skipper and crew of the trawler Maggie Comb saw the "immense monster" at about 200 yards distance. It seemed to be more than 120ft. [*36½ metres*] long, but no head or tail was visible. It lashed the water into foam, and then disappeared!

At first I wondered whether this might not have been a blue whale, its size greatly overestimated. Nevertheless, the distance was fairly close, and although blue whales can reach 30 metres, they would certainly appear much shorter if the head and tail were not visible. It is also unlikely they would remain invisible if it were lashing the water to foam.

Near Fraserburgh, Scotland, 1900

The following dramatic account should serve as a warning to leave well enough alone in the presence of large, unknown animals. The location appears to have been somewhere near the Rattray Head Lighthouse at 57° 36½' N, 1° 49' W, which is not far from Fraserburgh. The story comes from the *Telegraph* (Brisbane) of Saturday 1 December 1900, on page 3.

> Sea Serpent Again
> Aberdeen Trawler
>
> A terrible story is told by the captain and crew of the steam trawler Craig-Gowan, of Aberdeen, which arrived in Fraserburgh recently, storm-bound. Having heard that the crew of the Craig-Gowan had seen some strange animal when a mile or so north of Rattray Head, a correspondent of the Aberdeen *Journal* waited upon them and had an interview with the skipper, Captain J. Ballard.
>
> Captain Ballard said: "We left Aberdeen at 12 noon, and all went well, although the weather looked threatening, until off

Rattray Head, when the wind freshened almost into a gale, and the sea rose very rapidly. We were steaming 10½ knots [*19½ k.p.h.*] when the gale burst. At this time we noticed a smack some distance off seaward. She had the smallest bit of sail set, and was heading southward. This might have been about 4.30 p.m. or thereby. I went below for some coffee, and had been down a few minutes when J. Watt, chief engineer, called me up, saying that a whale or some extraordinary large animal had been following in our wake.

"On reaching the deck, I found several of the crew looking over the weather rail. On joining them I saw, greatly to my surprise, a very large animal of dark colour, which seemed racing with us, but which was about 50 feet [*15 m*] to windward. I had often seen whales, but I at once saw the animal was not a whale, but some sea monster, the like of which I had never seen in my life.

"As it rose, several portions of the body were visible at the one time. It seemed to make its way through the water, showing repeated portions of a dark brown body. The men seemed very much struck by its strange appearance, and I suggested to try some plan to get rid of it, and no one seemed to grasp any plan likely to affect so huge an animal. We had left our deck hose at Aberdeen, but I asked Mr. Watt and his assistant, Dallas, to bring up a furnace rake. The animal was now uncomfortably near. We could see that the skin was covered by some substance like a rough coating of hair. Securing the furnace rake to a stout line, I threw it at the animal, but it fell short. I again tried; this time the rake landed across the animal's back, and we suddenly drew the line.

"Judge our surprise and alarm when the monster raised its body (the fore part) clean out of the water, and made direct for the Craig-Gowan. Everyone rushed aft, some down the companion-way, and some down the engine-room stairs. I stood almost petrified with the sudden development of affairs. I plainly saw the monster rise up until its head was over our gaff peak, when it lowered itself with a motion as sudden as lightning, carrying away the peak halyards, and send the gaff, sail, and all down on deck. The utmost consternation amongst the crew ensued, and it was a time before we got matters squared up. The animal had then entirely disappeared, and we did not see it again.

"We held on for Fraserburgh, where we arrived at 6 p.m., and being afraid that our story would be discredited, we have said nothing about it; at least as little as possible, until this account, which is a true one, was given as stated."

The crew say they never even heard of such a monster, and that such a monster was until now quite unknown to any of them.

Mr. Ballard lives in Torry, where also Mr. J. Watt, engineer, resides. Dallas, Collie, Fraser, and Mackay are the other members of the crew, and reside in Aberdeen.

The Craig-Gowan has left for sea again, the weather moderating.

Captain Ballard describes this experience as being one of the strangest in his lifetime, and says he would not again like to undergo another quarter of an hour's terror like that gone through for any money.

"The animal's head," he added, "was long and flat, and I distinctly saw its eyes, and also saw its mouth open. Its body was long, and of a round shape on back, and flat below. Several large fin-like flippers played about rapidly, the sound of their flapping against the body being quite audible as it rose up out of the sea. It must have been of a great length; how long I cannot hazard an opinion."

An internet search reveals that a succession of trawlers named *Craig-Gowan* operated in the same area under the ownership of the Craig Gowan Steam Fishing Co. Ltd. One of them was wrecked in 1896. The ship involved in the above adventure was presumably the one built in 1897, with a length of 95 ft. 6 in. [*29.11 m*], and a depth of 10 ft. 2 in.

[*3.10 m*]. However, it is possible that it was its successor, from Yard No. A323, which had been constructed in 1899 under the name of *Cortes*, which was slightly bigger, and is shown on the photograph above.

On balance, however, I think the relevant ship was the 1896 version, because the 1899 version does not appear to carry sails. Nevertheless, it is not difficult to find pictures of steam trawlers of a slightly earlier date equipped with sails in addition to engines. A peak halyard is a line close to the top of a mast. Therefore, for the sea serpent to have reached it, it must have raised its head at least 20 feet [*6 metres*] and more like 30 feet [*9 metres*] out of the water, leaving who know what length of body in the sea. Its movement, showing repeated sections of its body, suggests vertical undulations which, in addition to the appearance of hair, indicates a mammal of some description.

Chapter 5
The *Tresco* Sea Serpent of 1903

On pages 368 - 369 of *In the Wake of the Sea-Serpents*, Bernard Heuvelmans describes the sea serpent allegedly seen by the crew of the *Tresco* near Cape Hatteras in 1903. His opinion was that the story "reaches the peak of fantasy", and I have to admit that, from the summary provided, it did seem to suffer from credibility problems. Nevertheless, one always wishes to refer to the original document, which was cited as the October 1903 issue of *The Wide World Magazine*. Now, *The Wide World* was a monthly magazine in which members of the public related their own adventures in various parts of the world. It was a requirement of publication that they certify that the story was true in all particulars and, in most cases, I suspect they were. Most of them lacked the normal structure of fiction - the beginning, middle, and end - and had the air of truth about them. Just the same, there was no method of confirmation, and a number of hoaxes certainly did make their appearances in its pages.

I have been an avid fan of *The Wide World* ever since I was introduced to it as a boy, and have collected every edition I could lay my hands on. Regrettably, this includes only a couple before World War II. Then, a while ago, a light bulb went off in my head. The Internet Archive contains a number of the early bound editions, including volume 12, where the relevant article appears on pages 147-155. I would really like to introduce you to this wonderful magazine, and I would seriously suggest that you read it[2]. But for those who lack either the time or the inclination, I shall publish the article here.

The "Sea-Serpent " of the "Tresco."
By Joseph Ostens Grey, Second Officer of the SS "Tresco."

Will the problem of the "sea-serpent" ever be satisfactorily solved? Scientists and others scoff at the idea of its existence, and cast ridicule upon those who claim to have seen it; nevertheless, hardly a year passes without a seemingly well authenticated account of its appearance being added to the cases on record. We publish herewith the story of Mr. J. O.

---

[2] It can be downloaded and read at
https://archive.org/details/wideworldmagazin12londuoft.

Grey, second officer of the SS "Tresco," of the well-known Earn Line, whose statements are corroborated by the captain of the vessel and other eye-witnesses.

Seafaring men expect storms and sometimes wrecks, but for most men of the merchant marine in times of peace there is much monotony in their voyages to and from the various ports they seek during their years at sea. On an ordinary voyage, such as I have taken, year in and year out, for sixteen years, a remarkable experience befell me recently.

I know that the very word "sea-serpent " is the signal for joking, ridicule, and utter incredulity. While many reports have been brought to land, no sea-serpent, small or large, and no fragment of head or fin have ever been subjected to study by any recognised scientist; and yet such a creature confronted the steamship *Tresco* when on her last outward voyage from the United States.

We left the port of Philadelphia, in Pennsylvania, on May 28th, 1903, for Santiago de Cuba, which we reached on June 5th, and we arrived back in Philadelphia on June 14th. The *Tresco* belongs to Mr. E. C. Thin, a shipowner whose office is at 27, Chapel Street, Liverpool; she is under a two years' charter to the Earn Line, of Philadelphia. The *Tresco* is a large cargo-steamer engaged in the West India trade. She plies from one port to another, usually laden with sugar, but sometimes with iron. Her length is three hundred and eight feet, her registered tonnage one thousand eight hundred and sixty tons, and her gross tonnage three thousand seven hundred and fifty tons.

On this trip it so happened that, instead of the *Tresco* being heavily laden with a return cargo, she was going out in water ballast; the ship was therefore very light. She rose well out of the water, her rail some twenty feet above it. Her draught was no more than twelve feet and she was extremely "tender." Twenty tons of coal deposited on either side of the main deck would have given her a dangerous list to port or starboard, as the case might be. We encountered no heavy weather and all went well on board; it was the true monotony of the merchant marine.

Our crew, of course, changes from trip to trip, but our officers have been a long time with the company, all of whose ships have somewhat similar names, beginning with Tr, like *Tripoli* and *Tronto*. Our skipper is Captain W. H. Bartlett, whose home address is James Villa, Looe, Cornwall ; our first officer is Mr. Elias Griffiths, who lives near High Park Street, Liverpool. I am the second officer - Joseph O. Grey. We had twenty men on board [*indecipherable*].

The next couple of paragraphs are impossible to read due to the text not copying well. However, he does state that, two days out, they were on an oily sea about ninety miles from Cape Hatteras.

About ten o'clock I saw, on our port bow, something creating a vast amount of disturbance in the water. The commotion was so great that I judged it to be a school of porpoises, which herd together and play, jumping above the water like great Newfoundland dogs. It is not at all uncommon to see a school of them in those waters ; but, somehow, the approaching school seemed different. I watched them closely as they neared the vessel from the south-east.

Whatever was approaching the vessel, the water was surging about some large fish which presently I discovered were not porpoises, but sharks. Now sharks are common enough, but not in solid masses as was the school I now beheld travelling at such great speed. It seemed to me a phenomenal departure from anything I had heretofore observed in regard to these voracious and savage creatures. They were not attracted to the vessel by anything thrown overboard, but held steadily on their way.

They seemed to be some maritime express, bound for Cape Hatteras; for, from the time we sighted them until they disappeared, they kept to their course, as if making all speed. What impelled them to travel at such a rate I could not imagine; nor could I offer any explanation for their assembly in such a solid mass.

Sharks differ in size and there are several varieties. So far as I could tell these were the usual bottle-nosed shark. They were swimming shoulder to shoulder, closely packed together, their dorsal fins cutting the water steadily. Occasionally their snouts appeared. It was a curious spectacle, and, while in no way alarmed, I watched them until they were out of sight. In all, as nearly as I could count them as they passed, their number was about forty.

I saw no more sharks. The time went by uneventfully. My mind reverted several times to that rushing herd of sea-tigers, and no reason for such swift, steady pursuit of an unchanging course occurred to me. My wonder rather increased than diminished.

The passing of the sharks had made me unusually on the alert. About an hour later I espied a fresh object in the water on our port bow. It was some distance away, due south-east — exactly the direction from which the sharks had appeared. It was floating low, and it looked black. I thought it must be a derelict — one of those wandering, drifting hulks, so desolate to see, so dangerous to encounter.

I instantly gave orders to the man at the wheel to steer for the derelict. The *Tresco* was steaming along due south; but now she swung gradually about until she was going exactly south-east. The sea was still calm and smooth. We sped easily on our way, with little said except, "It is a derelict; steer for her."

The man at the wheel beside me on the bridge thought so too as we headed for it, wondering how much of a hulk it would prove to be, or what we should ascertain of its history. We always steer for derelicts in the hope of possibly rescuing survivors; or some poor bodies may remain that need decent Christian consignment to the sea. It is, besides, an important duty resting upon the masters of all vessels to report to the Hydrographic Office the name of every derelict met with.

During the twenty minutes we were steering toward it I was decidedly puzzled. It seemed to me that this low-lying, dark object was moving toward us, as well as we toward it. It did not look like the hull of a vessel; nor could it be a raft. Neither would move so swiftly toward us. What could it be? The puzzle grew stranger. I stared intently, as every moment brought us nearer. We would soon know, at all events. The powerful engines were driving us onward so rapidly that the solution would be now a matter of but a few minutes. And yet the time seemed long. Nearer and nearer we drew and at last we were but two ships' lengths away. With a conviction that grew ever deeper, and ever more disquieting, we came to know that this thing could be no derelict, no object the hand of man had fashioned, no object, probably, the eyes of man had ever seen.

Now, swiftly, with a terrible uprising, a mighty and horrible head came out of the water, surmounting a tall, powerful neck that had the thickness and strength of a cathedral pillar, yet spindly in proportion to the huge and awful head it supported.

The next couple of paragraphs are also next to impossible to read, but they appear to relate to the absolute panic of the crew in the presence of "the dragon-like head and ... the long, powerful neck."

I felt that I must run somewhere, anywhere, to get away; and yet the weird and awful thing, there before us, held my gaze in the one direction.

At length I recovered some measure of my self-possession.

"Jump, Leon; jump down into the wheel-house!" I shouted. "Steer down there. Let's get out of this fellow's road!"

Magazine illustration of the monster.

The man obeyed with alacrity; and I, only too gladly, followed him. There were seven steps to be descended; and I felt like a child afraid of the dark does when it runs upstairs to bed, thinking a bogey is after it in the hallways. I was frightened; there is no use to deny the fact.

Once inside the wheel-house, I flung the door to and locked it, thankful for even this frail barrier— thankful for the slight protection of the wheel house, a mere nothing to such an adversary. There we were, silent both of us. Leon took his place at the wheel. We waited for what was to come next, still with the same sense of awe and huge, overwhelming dread upon us.

The wheel-house and chart-room adjoin, being one compartment with a partition. In front there are four windows, commanding a wide range; but, unluckily, from his position at the wheel Leon could no longer see the object. It was too near. He stayed at his post, needing no orders. I stepped into the chart-room to his left, where I could obtain a full view of the serpent as it faced us.

I could see it steadily and well from the chart-room port-hole. I looked and tried to notice every possible thing about it, yet wondering anxiously all the while how we should escape. The man at the wheel, and I with my face close to the port-hole, were stricken too dumb with astonishment and fear combined to say a word to each other. We did not say, "What is it? What shall we do if it comes nearer?" Nor did we discuss its appearance and actions. To me it was sickening and horrifying, and Leon had seen quite enough before he fled from the bridge.

Out of the formless horror within me a dread arose which shaped itself into a distinct, dismaying apprehension. What if the thing should attack the steamer ? The consequences loomed up, fearfully appalling, to my swiftly realizing imagination. The creature, assuredly, was enraged. So enormous was its size, so vast its strength, that even a steamer like the *Tresco* would be in danger of some kind — perhaps of many kinds. The rail of the ship, it was true, was twenty feet above the water ; but the head and neck of the serpent were already elevated to a height of fifteen feet. It could easily come aboard. The whole deck, all the upper works, in fact, would be at the mercy of its rage !

But far more serious to contemplate was the problem of its mere weight. That alone was a menace to the ship's safety. As I have said, we were going out in ballast, very light. Such a weight on one side would inevitably list the vessel, for the centre of gravity was so high that any heavy, ill-placed burden meant the gravest danger.

There that evil thing remained, the body motionless, the tail undulating vertically. As it lashed the water with the long, snake-like tail the head all the time was reared high, regarding the *Tresco* as if waiting to see what such a thing as a ship might be and, until it should decide, determined to maintain its watchful position. It looked for all the world like some fantastic Chinese dragon become a living reality; or a page from a scientific work picturing some ancient saurian monster, neither reptile nor beast wholly, but both in part.

When I first saw it, lying so low as to appear like a derelict, I must have seen only the back and body. The head was probably resting on the shoulders, as a swan sometimes rests, until, coming within two ships' lengths, we alarmed it by our unfaltering approach to the position of defensive attention.

We needed no binoculars. A sailor sees as no landsman sees; his eyes are trained to watch sky and sea and every object which may affect the welfare of the ship. And, indeed, the serpent was so near that even untrained eyes could have distinguished the most minute details of its appearance.

I estimated the length of the creature at about one-third that of the *Tresco*, or one hundred feet. We saw it only in perspective up to this time, for it remained facing us, neither wheeling nor changing position.

I judged it to be about eight feet in diameter in the widest part of its body, and so about twenty feet in circumference. The body was not cylindrical at all. It had a noticeable arch toward the top, and the hump of the back sloped downwards to the neck as well as toward the tail. It was widest at the forward end, rapidly tapering backward from the hump above the shoulders.

There was something unspeakably loathsome about the head, which was five feet long from nose to upper extremity. Such a head I never saw on any denizen of the sea. The neck, eighteen inches in diameter, was slender by comparison. Underneath the jaw there seemed to be a sort of pouch, or drooping skin; there may have been a slight bulge there. The neck was smallest half-way between the head and where it joined the body.

The nose, like a snout upturned, was somewhat recurved. It was rather pointed in its general formation, but blunt at the end. I can remember no nostrils or blow-holes. The lower jaw was prognathous, and the lower lip was half projecting, half pendulous. Presently I noticed something dripping from the ugly lower jaw. Watching, I saw that it was saliva, of a dirty drab colour, which dropped from the corners of the mouth.

While it displayed no teeth, it did possess very long and formidable molars. There were two and they curved backward like walrus' tusks [*indecipherable*] inches in length, at the [*indecipherable*] of the mouth. They were of a dirty [*indecipherable*] colour. If it had teeth or tongue it did not [*indecipherable*] mouth was red.

Its eyes were of a decided reddish colour. They were set high in the head, like [*indecipherable*] water-fowl. [*indecipherable*] They were elongated vertically. [*indecipherable*] They carried in their dull depths a sombre, baleful glow, as if within them was concentrated all the fierce menacing spirit that raged in the huge bulk behind.

Below the eyes some scales appeared, which dragged backward, becoming larger and larger until, on the body, they were great plates, or protuberances like the denticulated ridges of an alligator's hide. They did not glisten like the scales of a fish. The smallest of the scales, near the eyes, measured about three inches in diameter, and were so little oval as to appear completely round. The largest of the scales, or indurations, located upon the shoulders, presented a form more pronouncedly oval, and these were some eight inches long, five inches wide, and four inches high, their apex being a distinct ridge.

The hide, in the general tone of its colour, could be compared to nothing but antique bronze, showing the distinct light green hue of the oxidized metal. The tone of the colour was lightest upon the back and sides. As it shaded toward the almost wholly submerged belly it became a dull, dark green, deepening its hue with the decrease in the size of the plates or indurations constituting the creature's defensive armour.

It held itself in the same relative position to the ship during all the time the impressions I have enumerated were photographed indelibly on my brain. Its side fins, extending one-third of the way from the shoulder to the beginning of the tail, and broadest — about a foot — near the shoulder, worked like fans in swift agitation of the water.

As I gazed, fascinated with the horror of the thing, it raised its dorsal fin, obviously in wrath. And then a thing happened which, strange as it may appear after the recounting of the fearsomeness of the serpent's dreadful front, was more appalling, more sickeningly terrifying, than anything I had yet beheld. Suddenly, at the back of the head, a great webbed crest uprose, and from the eyes, hitherto so dull save for the glow smouldering in their depths, a scintillating glare appeared, as if the creature felt the moment had come for attack. The crest was a foot in height at its forward extremity, where it was supported by a sharp-pointed spine.

The undulations of its tail increased in violence. It lashed the water in fury. Its reddish eyes were fixed upon us; but, threatening as it appeared, it came no nearer. The novelty of our appearance, and our size, seemed to make it hesitate. In what way it would have attacked us I can only imagine.

This hesitation and anger, combined, kept it at a standstill, and our fear and helplessness for resistance kept us quiet. The creature remained in this fashion, glaring at us, for a few moments more. Then I saw it was about to act.

It was going to turn away from us. I could scarcely credit my senses. I watched its new tactics carefully. Yes, it was moving and turning; it was about to go from us. I felt an infinite, deep-breathed sense of relief.

Its great body turned, as if on a pivot, inward in a circle, followed by its long tail. With astonishing ease for so huge a bulk it made the sweeping evolution. And only then did it lower its ugly head, that had so long confronted us in open antagonism. I began to breathe more steadily. I was certain now. It was afraid, and would go peaceably.

Only at that last moment did I think of Captain Bartlett. I must call him, now that I dared to venture out. I wanted him to see the monster. I unlocked the door and flung it wide, and ran aft along the starboard side as fast as I could. I burst in upon the captain in his state-room. He was lying down, but was fully dressed. The noise of my entrance startled him.

"Come on, captain, quick!" I exclaimed. "Come up and see this animal!"

Springing up instantly he was ready to follow. He comprehended that something unusual was near, yet he was astonished at such a report from an excited mate, five seconds more and we two stood together on the poop, where we could have a clear view and, as I knew now, a safe place from which to gaze upon our gruesome visitant. I was half glad, half worried to find it was still in sight. The captain would not think me demented.

Captain Bartlett stood transfixed. A moment and he found his voice: "Good heavens ! What's that?"

"I take it, sir," I replied, "to be a sea- serpent."

"I believe you're right," he rejoined. We stood there waiting to see whether it would go or return.

The serpent, or whatever else it may have been, was on our port quarter, for the engines had been driving us steadily ahead. The distance at which it was then removed was about a quarter of a mile. Its tail was now toward us. The back of its head, sunk upon the shoulders, was visible, together with the twenty-five feet of the body which I have hitherto characterized as the hump of the back. As we [*more indecipherable paragraphs. However, he appeared to liken the monster's submergence to the sinking of a water-logged wreck bow first.*]

[T]he terrifying thing was gone we could talk and compare our observations and ideas concerning it.

As I have said, I did not notice any nostrils; but I believe it was a breathing animal, endowed with lungs. While no sound reached my ears as we approached it, and while Leon and I were hidden in the chart-room, Captain Bartlett thought he heard distinctly, as we stood side by side on the poop, a noise which came from the creature that was in the nature of a snort or, to be exactly correct, a hoot. The sound, according to the recollection of Captain Bartlett, might be compared to the noise of a shrill tug-boat whistle. For myself, I must frankly say I can recall absolutely no sound. The coincidence of the appearance of the sharks and of the great lizard during the same hour is something I can affirm but cannot attempt to explain. An inference that would seem obvious is that the sharks were fleeing from the monster. But, in the absence of definite knowledge, it must remain coincidence, and nothing more.

After the exchange of these few observations Captain Bartlett turned to me and said : —

"I have had many strange experiences, as you know; and I have seen many strange sights. But I confess this thing is, without doubt, the most horrible and blood-curdling that I have ever looked on. Grey," he continued, "words cannot describe its loathsomeness, or the horror and terror with which I gazed upon it."

All this time none of the crew had dared come on deck. Our chief officer, Mr. Griffiths, was asleep in his cabin. The men who had fled so hastily, and the others who came at their call, looked out fearfully at the serpent from the forecastle ports. The steward, John Jackson, a coloured man from Baltimore, who saw it, was greatly terrified. He has since left the *Tresco*, having been engaged only for the voyage. Those who did not see it, like Chief-officer Griffiths, can testify to the general excitement and the facts elicited by the subsequent discussion among the men who did.

When the danger was over the men cautiously returned to the deck. Faces appeared at the hatches, and, after a little reconnoitring, up the companion-way they came, looking carefully astern, to assure themselves that the monster was really gone. Gradually, as they regained courage, they resumed their work, although they were careful to remain in groups, still talking over the astonishing event. After a long time had elapsed they were hardy enough to joke about it, although they had been so scared; and they repeated the story to the men in the engine-room, who had, of course, not even caught a glimpse of the stranger.

All this time the sea had remained quiet and the weather the same, so the conditions throughout were most favourable for viewing the monster.

I now ordered the vessel to be put on her course again — due south. The incident was over; our work was before us. Whatever danger had existed was passed. Santiago was to be reached, and we made that port on the fifth day afterward. As I watched through the port and, later, on the bridge, when, my fear abating, I could collect my thoughts better, I wished we possessed powerful guns which could tear a hole in that appalling head or through the armoured body, so that we could secure the carcass as a trophy and settle once for all the controversy concerning the sea-serpent. And I clenched my hands with annoyance, as I have clenched them many times since, when I thought of that camera of mine, ashore and useless, awaiting my next trip to St. Thomas. Why had I left it there, when now, for the first time in my life, I really needed it ?

During the five days that were required for the remainder of the voyage our conversation naturally reverted to the exciting morning and to the experience we never expect will be ours again. I, for one, sincerely hope it will not be repeated, unless for the corroboration of this statement and to assist science by delivering to some learned body the carcass of another such monster.

We have carefully collated all the facts. Our conclusion is that the creature was, without doubt, a mammal, like porpoises and whales, although more like a reptile in appearance.

At Santiago I prepared a report for the Press of Philadelphia, to be presented on my return. Although I made it out carefully, it drew forth the usual jests in several quarters, but it was credited in others. How bitterly I have regretted that I had no photographs to settle the doubts of those who questioned the accuracy of the drawings I have since made from memory ! I have but to shut my eyes, and that ineffaceable picture rises before my mind in all its horrible detail.

So, What's Wrong With This Story?

Well, we could start with the tone. Heuvelmans complained that the crew's panic was "described in such terms as Poe or Ambrose Bierce might have used." Well, yes. But here an explanation is due. It is a sad fact of life that literary skill and the sort of experiences worth writing about are gifts only occasionally granted to the same person - an anomaly which provides employment for ghost writers, but nevertheless deprives the world of a lot of worthwhile stories. Typically, articles for *The Wide World* were rewritten by the editor. One trusts that he merely improved the style and adjusted the length to fit the pages, but did not change the substance. But this article contains more purple prose than was typical for the magazine which, what with the newspaper report to be discussed below, suggests to me that it was in the original.

Another thing to understand is the artwork. Whether it involves sea serpents, UFOs, or any other anomaly, typically the artwork takes on a

life of its own, and is presented much more frequently than the text. It is also important, therefore, to ask yourself whether it was produced by:
- the witness, who may or may not have any artistic skill;
- a trained artist under instruction from the witness (probably better); or
- a trained artist illustrating the witness' account. This can be misleading. I myself once noted that the drawing of the *Umfuli* sea serpent failed to match the text. R.T. Gould pointed out that the drawing of the famous *Daedalus* sea serpent was foreshortened to fit the pages of the *Illustrated London News*. Darren Naish has suggested that the illustration of the questionable *U28* monster was probably done by somebody looking at a baby crocodile preserved in alcohol. As for the oft-reproduced sketch of the monster seen by Hans Egede in 1734, this was was produced many, many years after the event.

With respect to the *Tresco* monster, the drawings (there were more than one) are good enough as far as they go, but they were made by the magazine illustrator without any input from the writer - unless they were based on the "drawings" mentioned in the last paragraph.

Heuvelmans also questioned the extreme precision in the perceived dimensions of the scales: "some eight inches long, five inches wide, and four inches high" - and rightly so.

Problems abound in the described anatomy of the monster. For a start, it was, to put it bluntly, a tad on the large size. Just the same, Heuvelmans recorded many sea serpents as large as the biggest whales, and fear can make one overestimate size, so perhaps we can leave it as that.

Despite the author's assumptions, it must have been a reptile rather than a mammal; mammals are not covered with scales, nor do the possess dorsal fins or crests on the head. However, they do move with vertical undulations. A reptile would lash its tail horizontally, rather than vertically. (In the vast majority of sea serpent sightings, the tail has not been visible.) But what can be made of the fins - a foot wide, extending along the side of the body, and beating like fans? A reptile or mammal would have paddles, not fins. There is something fishy there. Even so, they sound too narrow to manoeuvre an animal of that size. Also, I wonder why on earth its colours would be darker on the belly than the back, exactly the reverse of nearly every other swimmer. It would make it stand out very clearly when viewed from below. And let's not forget that every huge monster was once a little monster, vulnerable to predators.

Finally, there have been several hundred sea serpents recorded over the years. Some of them possessed what might be considered scales. But a head with a pouch, two walrus-like fangs, and a crest, not mention "baleful" red eyes, is unique. When an alleged sighting is of something biologically unlikely, and is never confirmed by a second report, the assumption must remain that it is untrue.

On the Other Hand . . .

Heuvelmans did go to the trouble of checking *Lloyd's Register*, and confirmed that the author was pretty accurate with respect to the *Tresco*, which caused him to add: "The writer clearly took more trouble with his shipping information than his zoology." But there there is more to it than that. I would strongly recommend that you follow the link I provided and read the original PDF article, because it contains photographs too numerous to copy here: of the ship, the author, the captain, the other witnesses, and the author and quartermaster in the exact positions occupied when the monster was first sighted.

It also includes a sketch of the course of the ship, and copies of three documents. The first was the author's certificate that the account was completely true, and dated 22 June from Philadelphia, which was presumably the date the article was sent to *The Wide World*. (You will remember that they arrived back at Philadelphia on 14th.) The second was the enclosed certificate from the captain and the other witnesses, their signatures certified by Chief Officer Griffiths who, you will remember, did not see the monster himself.

Most important of all was a fascimile of the ship's log for the date in question. As you can see, it was signed by Chief Officer Griffiths, and contained the citation:

10 am passed school of sharks followed by a huge sea monster.

FACSIMILE OF A PAGE IN THE "TRESCO'S" LOG SHOWING THE ENTRY CONCERNING THE SCHOOL OF SHARKS AND THE SEA MONSTER.

Now, falsifying a log is a pretty serious offence, not something one would rush into just to satisfy what was basically a lark by the Captain and Second Officer - especially since it was not exactly essential for the hoax. You will note, too, that the home addresses of the Captain, Chief Officer, and owner were all provided, possibly under the assumption that someone might want to check up on it. There was also a small titbit which a hoaxer would be unlikely to add: that the Captain heard a "snort", which the writer didn't.

In addition, there was the last paragraph, about preparing a report for the Philadelphia press. I don't know how the Philadelphia press took it (perhaps one of my American readers could look it up), but *The Chicago Daily Tribune* was more obliging, publishing this paragraph on page 12 of its edition of 17 June.

THE SEA SERPENT OF 1903.

The sea serpent of 1903 at last has put in its appearance. It was observed by truthful Capt. Bartlett of the good steamer Tresco. When ninety miles south of Cape Hatteras the captain noticed a peculiar disturbance in the distance and that the disturbance was headed his way. In a short time it arrived and turned out to be a great number of sharks tearing through the water "like all possessed " on their way towards shore. An hour afterwards Capt. Bartlett noticed a dark object in the distance, and, thinking it a derelict, steered for it. When a short distance from it the "derelict" slowly lifted itself above the water. Instantly the crew went below and barricaded the doors, and through the port-hole in the wheelhouse the captain saw the sea serpent. "Half dragon and half serpent, it was the most hideous and loathsome reptile, with its gaping jaws and bloodshot eyes. From each side of its horrible mouth two large tusks protruded, similar to those of a walrus, and its lips were dripping with a discolored saliva which emitted a most offensive smell." The captain was not in a condition of mind to make accurate estimates of measurement, but he thinks the serpent was about 100 feet long. Fortunately, instead of making an attack "it turned tail, and, with a swish and a swirl of the water, sank in the depths," probably much to the relief of Capt. Bartlett and his crew, not to mention the panic stricken bunch of sharks. Now, this is a sea serpent worth having. The veracious captain's statement shows it has not suffered "a sea change into something rich and strange," but retains all its old horrible and terrifying aspects. It is no ordinary cephalopod but a worthy descendent of the serpent which Regulus and his army made war on with catapults, and of the monster Olaus Magnus saw, which not only ate calves, sheep, and swine but also "disturbs ships, rising up like a mast and sometimes snaps men from the deck," or even of Fafnir, spurting smoke from his cavern. The sea serpent has been an unconscionable time putting in an appearance, but better late than never. For what should we do in these days of strikes, and floods, and cyclones, and cloudburst, and droughts, and fires, and barbaric regicides had we not our old friend the sea serpent for pleasurable entertainment ?

You will note that here the Captain becomes the chief protagonist, rather than Joseph Grey. Perhaps he did that on purpose, or perhaps there was some confusion. (I have enough experience of the press to know that they usually get some detail wrong.) Perhaps he said "we", and the journalist assumed it meant "I". Perhaps he simply handed over Grey's report, because the newspaper appears to quote directly from it. (And note the "purple prose".) It is always possible that *The Chicago Tribune* simply picked up the story from a correspondent in Philadelphia. In any case, it appears that the Captain went out of his way to present the story upon landing.

If this was a hoax, it was a pretty elaborate one, involving several people.

I am reluctantly driven to the contradictory conclusions that it was not a hoax, but that the animal reported could not have existed. Don't you just hate it when that sort of thing happens?

Chapter 6
Edwardian Sea Serpents

The Edwardian period is, of course, the years marked by the reign of Edward VII, from 1901 to 1910. However, it is more convenient to stretch it out to the start of the First World War, which marked the end of an era.

In these early years of last century reports of sea serpents continued hot and strong. The fall off in later decades was not due to any increased rarity of the animals, but merely that I am concerned only with *forgotten* cases ie those which appeared in Australian newspapers, but which remained unnoticed by later researchers, especially the indefatigable Dr. Heuvelmans. Indeed, most of those he missed were specifically Australian sea serpents, and were published solely in Australian newspapers - as I have cataloged in my earlier work, *Australian Sea Serpents*. However, foreign sightings were originally published in overseas journals, and those Dr. Heuvelmans was able to access. Even so, I have noticed that after the Second World War newspapers no longer reported them, or else witnesses were less inclined to come forward.

Near Hong Kong, 1901.

This time we have a very detailed account, published in the *Express and Telegraph* (Adelaide, SA), Wednesday 6 November 1901, on page 3. The approximate site, Boddam Cove, lies on the northeast side of Tungho Island, at 22° 02' N, 113° 43' E.

THE SEA SERPENT
SIGHTED IN CHINESE WATERS
(From our Special Correspondent.)
London, October 4, 1901

Our old friend the sea serpent has made his appearance again, this time in troubled Chinese waters. The monster, whose existence there appears no reason to doubt, has been the subject of an official report by Mr. Wolfe, who has been in the Chinese Maritime Customs service for nine years, and is in charge of the armed revenue launch Lungting, a vessel of 100 tons, with a speed of 14 knots [*26 kph*]. Mr. Wolfe is certified by one of the Customs authorities to be a steady, trustworthy, and credible man, and his evidence is confirmed by his second officer and all the Chinese crew. Here is Mr. Wolfe's narrative, which speaks for itself:-

"On Sunday, August 18, 1901, at 11.20 a.m., as the launch Lungting was steaming at half-speed, heading for Boddam Cove, Tungho Island, at about 10 cables' length [*1.85 km*] from the Chuk Chao Island, I sighted a dark object on the surface of the water one point on the starboard bow, which looked to me like a rock. I at once gave an order 'Full speed astern,' and the vessel passed about 30 ft. [*9 m*] clear of the object, which, to my surprise, was a large serpent, lying in a round coil, with its head raised 2 or 3 ft. [*60 or 90 cm*], and slightly moving. Stopped engines and lowered starboard gig. I dispatched Mr. Kuster, second officer, in gig, with orders to kill the monster, if possible. Mr. Kuster stood in the bow of the gig with a boathook ready to strike. The serpent had now lowered its head again, but on approach of the gig suddenly struck out, hitting the blade of one of the oars, turning the sailor turtle-back. It then raised its head to a level of the launch davit, about 15 ft. [*4½ m*], at a distance of not more than 10 ft. [*3 m*] from the gig and 30 ft. from the launch, where I stood. The crew of the gig were scared, and prepared to jumped overboard. Mr. Kuster, still standing in the bow of the gig, prepared to strike with the boathook; but, before he could do so, the monster suddenly dived, and made off. Its actions in swimming was like that of an ordinary water-snake; the water being clear, the reptile could be plainly seen a few feet down. It dived very quickly, and made considerable disturbance of the water. We judged the serpent to be from 40 to 50 ft. [*12 to 15 m*] long and about a foot [*30 cm*] in diameter. It had a kind of crest on its head, and two fins high up on the neck, just behind the jaws. The thickest part of its body appeared to be about 15 ft. [*4½ m*] from the head, tapering both ways. Its head was as big as a Rugby football, with large eyes, and mouth opened wide when striking. It was of a very dark color on the back - striped and mottled, but lighter on the belly. As soon as the serpent disappeared, and we on the launch had recovered from our first surprise, I ordered the ten-barrelled Nordenfeldt to be loaded and the launch moved round slowly for 15 to 20 minutes, in hopes that the reptile would reappear; but not doing so, the vessel proceeded on her way to Boddam Cove.

- (Signed) F. Wolfe, officer in charge C. L. Lungting, August 21, 1901. Witnesses - (Signed) V. Kuster, second officer, and 17 Chinese.

This is very strange. At first glance, it sounds like a genuine snake, despite the huge size, but the presence of two fins "high up on the neck" suggests some sort of eel. Similar sightings are rare, but have been recorded.

Indian Ocean, off South Africa, 1901.

This appears to have been a very interesting voyage. The story comes from the *Evening Telegraph* (Charters Towers, Qld), of Tuesday 10 December 1901, on page 2.

A NAUTICAL EXPERIENCE

The steamer Heathdene, which has arrived at Wellington (N.Z.) from New York, had a sensational experience on the voyage out. A fire broke out in a part of the vessel close to where 18,000 cases of petroleum were stored. After great effort on the part of the officers and crew, the flames were suppressed, but not before the steamer had been in imminent peril, for a wooden partition separating the petroleum from the spot where the fire was raging had been actually burned through. Whilst the captain and his men were below fighting the flames, the captain's wife (Mrs. Milburn), and her daughter steered the vessel, one keeping a look-out whilst the other controlled the helm. Miss Milburn (the captain's daughter), who is only 14 years of age, is said to be able to steer a ship as capably as any sailor. As if the fire on board did not comprise sensational incidents sufficient for one voyage, the crew of the steamer report also that when she was 10 miles [*16 km*] off Natal, an extraordinary sea monster was discovered, presumably the sea serpent again. This creature, which was of abnormal size, had a white fin, rising 10ft. [*3 m*] high from the water, and a long white streak running the length of its body, which otherwise was black in colour. An officer of the steamer, who has a fairly extensive knowledge of the varied species of fish in the ocean, declares that he has never seen anything like this monster.

Vancouver, Canada, 1902.

As mentioned before, right up to the present day, the coast off British Columbia and Oregon has been notorious for sightings of a mystery animal given the whimsical name of "Caborosaurus". However, it has normally been described as a long, sinuous body, not something towering in a vertical column like this one. The account comes from the *World's News* (Sydney) of Saturday 20 September 1902. Note that the story took 6½ weeks to arrive in Australia.

THE SEA SERPENT AGAIN
SEEN OFF VANCOUVER

A Vancouver, B.C., dispatch of August 5 says:

- "The best sea serpent story that has been developed on the coast in years was brought to Vancouver to-day by the fishing steamer New England. A distinctive feature is that not one but 15 men claim to have seen the serpent for five minutes at a stretch, and these fishermen are willing to make affidavit that their statement is absolutely true.

"The sea serpent incident occurred on Saturday [ie *2 August*], off the northern end of Vancouver Island. The fishermen had gone out in the morning, and were at different distances from the steamer. There were many whales spouting around in the vicinity, and the halibut catch was large.

"All at once an object arose out of the water a little to one side of us," said Alexander Easler, in describing the incident this morning. "I paid no attention at first, as we were busy in pulling in halibut, until my partner drew my attention to it. The fish, or whatever it was, pulled itself 30ft. [*9 m*] out of the water, and was almost as straight as if if had been a fixed column in the water. There must have been at least twice the same length under the water to support the immense weight of the body in the air. The fish moved at right angles to us, and left a distinct wake behind. It was very near, not more than a hundred yards away I should think, and the steamer was quite a distance away.

"I called to my partner to look out, and he stood by to cut the gear clear from the boat, so that we could get away if the thing came towards us. It was in the air four or five minutes. and then gradually went out of sight. I have been to sea for 30 years, and I never saw anything like it before. We did not see the head plainly enough to tell what kind of mouth or eyes it had."

It was a pity they were not able to provide more detail - even so much as the colour, or the thickness. In any case, it does appear to have been one of the "long necked" variety, but *huge* - even if the height was overestimated. Of course, it could not have been a fish. There is a common view that long necked sea serpents are mammals. In that case, it should be pointed out that, with only a few exceptions, mammals possess only seven neck vertebrae. There appears to be some sort of genetic restriction. Even a giraffe has only seven neck vertebrae. The implication is obvious: a mammalian neck 20 or 30 feet long would be too stiff to manoeuvre effectively in water.

Sardinia, 1902.

I strongly suspect that this is a hoax, but it might as well be published, and left to the discernment of the reader. Here, Downunder, it was first picked up by an obscure rural journal, the *Euroa Advertiser* (Vic.) on Friday 7 November 1902, at page 2, and then exactly a week later in an equally obscure Victorian newspaper, the *Horsham Times*, also on page 2.

SEA SERPENTS

The sea serpent is with us once more (says an Italian paper), in fact it generally comes to the surface at this season of the year. If we may believe what is stated, there is a remarkable creature disporting himself in the Adriatic, accompanied by all his family of eight members - sex not stated. A boatman named Bonifacino, sailing in a small boat with mail bags from one place to another in Sardinia, found himself suddenly in the midst of this terrible family party. There were three other persons in the boat at the time, and each of these is ready to attest the truth of Bonifacino's statement. As far as could be seen the creatures were fully 70 ft. [*21.34 m*] in length, with eyes about 7in. [*17.8 cm*] in diameter. They swam in a vertical position, raising their heads high above the water. Flaps hung over their mouth like ordinary doors, cavernous-sounding snorts were emitted from huge nostrils, and water spouted from deep cavities on the tops of their heads. They appear, however, to have been quite harmless, as the boatman and his passengers were able to part company with them without sustaining any injury. This seems to be the only occasion on which the sea serpent has brought his family out for an airing, which of itself is interesting; but an equally interesting ichthyological phenomenon is the selection of autumn by these queer fishes, or marine reptiles, as the only season in which they deign to reveal themselves to human gaze.

It is, of course, perfectly correct that this is the only account of a school of sea serpents. All other reports indicate that they are solitary beasts. Likewise, spouting water from the head is, to say the least, a decidedly rare behaviour. The article also appears to be a paraphrase of a report in an Italian newspaper - and not a very good one, I would suggest. For a start, Sardinia is not in the Adriatic Sea; it is on the other side of Italy. Secondly, in place of my usual custom of making approximate translations of imperial measures into metric, I have cited them exactly to highlight another anomaly: Italians use the metric system. Surely an Italian boatman would have cited his estimates in round numbers, such as 20 metres and 20 centimetres?

1903 was a good year for sea serpents. Heuvelmans, in his classic compendium, listed a dozen cases for that year. However, here are a few which he missed. The first is rather strange, even by sea serpent standards.

Off Brazil, 1903 (?).

It is always useful to check the co-ordinates in an atlas, because some hoaxes have been known to provide latitude and longitude co-ordinates for land, just to see if they can get it past the proof-readers. This location of this one appears to have been just off the coast of Brazil. The report comes from *World's News* (Sydney) Saturday, 28 February 1903, at page 7.

A MARINE MONSTER
A MONSTER OF THE DEEP: THE "SEA SERPENT" SEEN IN THE SOUTH ATLANTIC.

Captain W. E. Staveley, of the steamship Clumberhall, sends the "Daily Graphic" a sketch of "a large sea monster, sighted in the South Atlantic, in latitude 21deg. 39 min. south, longitude 40deg. 12 min. west." The vessel was on a journey from San Francisco to London, and the following entry relating to the event was made at the time in the captain's private log: - "At 11.20 a.m. my attention was called, by the second officer, to a large sea monster, of an oblong and whale-back shape. In part it was of a light sand color, and partly grey, with large blotches of black. The monster measured approximately 150ft. to 180ft. [*45 to 55 metres*] in length, the highest part of the body to sea base measuring from about 15ft. to 20ft [*4½ to 6 m*]. It was in motion, and for at least five miles [*8 km*] in its wake the water kept perfectly smooth, as if something of an oily nature were issuing from its body. When first seen it appeared to be heading towards the steamer, and when within a cable's length [*185 m*] of her it headed away in a south-westerly direction. The head I could not make out distinctly, as it would only now and again show the extremes of the body." A few days after seeing this monster the captain was shown a copy of the "Newcastle (Eng.) Chronicle," containing the account of a large sea monster which had been seen in the neighborhood of Cape Breton, Nova Scotia, and, as the description tallied, he suggests that it may have been the same monster.

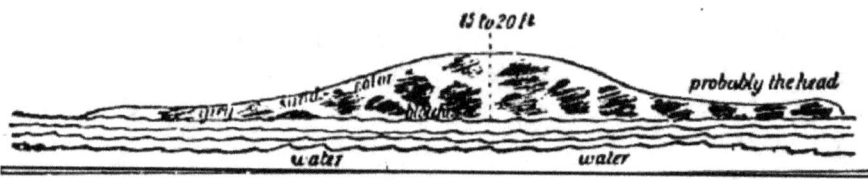

A MONSTER OF THE DEEP: THE "SEA SERPENT" SEEN IN THE SOUTH ATLANTIC.

Since making a false entry in a log is an offence, it would seem unlikely that the captain would have gone to that extent just for a joke. The story is thus probably genuine. But the creature was huge! Admittedly, its approach was no closer than a cable's length, but the witness was probably using a telescope. Just the same, how did he estimate the size? But even if it were overestimated by a factor of two (unlikely), it was still very big. Also, although many sea serpent sightings suggest a large body behind the neck, this is the first time I have heard of one lying stretched out on the surface, with its body riding so high. Also, what was its means of propulsion? It does not appear to have been by undulations.

As for the Cape Breton monster, Andrew J. Hebda has produced a magnificent ebook entitled, *The Sea Serpent Chronologies*[3], copying reports of sea monsters sighted off Nova Scotia (in a similar format to this book, only restricted to a particular site.) The incident referred to would be entry for 9 July 1902 on page 81, for the animal was estimated to have been 200 feet long, although only 50 feet of it was visible.

---

[3] It can be downloaded at
https://ojs.library.dal.ca/NSM/article/view/6411/5659.

Fiji, 1903.

This story turned up, more or less verbatim, in a couple of capital city dailies, the earliest one being *The Daily Telegraph* (Sydney) of Saturday 6 June 1903, on page 13.

The Sea Serpent Again

The sea-serpent has turned up again, this time off the Fiji group. The story runs that on the afternoon of March 13 last, a Fijian missionary named Meli, whilst on a visit to the lighthouse, had his attention drawn by Abdurahim, the Indian lighthouse keeper, to a large black object that was seen lying on the surface of the water close by the shore reef, and which is distance about 500 yards from the bluff on which the lighthouse is erected. From the heights of the bluff a good view is obtained, and the lighthouse keeper and his companion, the only two observers, were afforded the opportunity of witnessing the movements of the monster, which lay, as it were, almost under their eyes.

They described the strange visitor as being about 30ft. [*9 metres*] long, and about as thick through as a small cask, and showing nothing but a long snake-like body.

The creature lay quite still for a considerable time, then slowly raised its head out of the water to a height of about six feet [*1.8 metres*], and, apparently satisfied with the survey of its surroundings, started off with a kind of corkscrew twist, and with a big splash disappeared below, head first. The by now excited watchers of the unique spectacle pulled themselves together before the waters were again disturbed, and the sportive serpent once more came into view, and a similar performance to the first one was gone through. These evolutions were repeated several times, clearly demonstrating that the marine visitor was, in a way, thoroughly in accord with the theme of "A Life on the Ocean Wave."

Feeling satisfied that the aquatic display was over, the Fijian missionary betook himself for a stroll. His story concerning that stroll is to the effect that upon reaching the beach he saw the sea serpent lying on the beach, three parts in the water, with its head resting on the shore and looking and probably feeling as comfortable as could be. He describes the head of the creature as being about the size of a powder keg.

One really wishes they had provided a more detailed description, because the length and thickness don't sound too snake-like to me. They are consistent with some species of cetacean ie a whale or dolphin. The behaviour is also that of a cetacean, provided that we assume that there was no bend in its body when it raised its head out of the water. However, lying with its head on the shore is definitely not a cetacean behaviour. A whale or dolphin in that position would be stranded. For that matter, I don't know of any case where it has been recorded of a sea serpent, either, but I note Heuvelmans' opinion as to why there are no sea serpent strandings: a long, serpentine creature could wriggle out of the shallows.

Off Victoria, British Columbia, 1903
Of course, this is our old friend, "Caborosaurus", which has been raising its horse-like head off the coasts of British Columbia and Washington right up to the present day. This report comes from the *Australian Star* (Sydney) of Thursday 25 June 1903, on page 3.

SEA SERPENT VISITS INDIANS
A Victoria despatch dated April 14 says:-
Officials of the cable station at Bamfield Creek are sponsors of the story that a sea-serpent from forty to sixty feet [*12 to 18 metres*] long, and with a head like a horse, has been seen off the station. The cable operators say, in letters to the local newspapers, that Indians had been telling of the existence of a sea serpent, but the stories were not credited. David Osborne, one of the officials, say that a week ago the animal was seen from the cable station to raise its big, horse-like head and swim out of the mouth of the Bamfield Creek into Barclay Sound. Mr. Godson, of the cable staff, says that when he first saw the animal it looked like a massive sea weed, but presently he saw the head elevated and the big serpent move off with the speed of a torpedo-boat. On April 10 an Indian saw the thing, and was so frightened that he ran his canoe into the breakers, left it, and fled along the beach to the cable station. The Indian said the thing had a head shaped like a horse, and its body, ten feet [*3 m*] of which was lifted, was the size of a barrel. The Indians in the neighbourhood are terrified.

Montrose, Scotland, 1903.

Montrose, of course, is a city on the east coast of Scotland, just 38 miles or 61 km north of Dundee. This report comes from the *Catholic Press* (Sydney) of Thursday 29 October 1903, on page 23. You will note how the original journalist did not bother to ask any questions, or obtain any but the bare details of the story, except that, exaggerated or not, the creature must have been enormous.

Our Friend the Sea Serpent

This year the sea serpent has been long in coming, but the faithful creature has appeared at last. According to the British Press Agency, a trawler in operation off Montrose made out an extraordinary monster recently. All the hands have been at sea for years, and they have never witnessed anything like it before. The monster was swimming northerly at a great rate, and noiselessly. Its head appeared from four to six feet [*1.2 to 1.8 m*] long, and then 60 feet [*18 m*] behind was a huge protrusion, and behind again another protrusion. The monster's noiseless motion is said to have a most uncanny effect.

Orkneys, 1903

This brief article comes from the *World's News* (Sydney) of Saturday 19 December 1903, on page 20.

THE SEA SERPENT AGAIN

Perhaps disturbed by the magnetic storm, the sea serpent has made a tardy reappearance near the Orkney Islands and off the east coast of Aberdeenshire. Its only claim to novelty appears to be that in color it is "a pure white."

On Sunday morning (November 1), says the "Daily Mail", the fishermen at S. Andrews saw a strange sea monster, swimming eastwards, at the back of the old castle. It lived up to the best traditions by swimming with an undulating motion and exposing about 12ft. [*3.6 m*] of body at a time, it being about 4ft. [*1.2 m*] broad. Unfortunately, the head was modestly kept under the surface, and the full length of the serpent can only be guessed.

It was followed by a great number of seagulls and by two fishermen in a boat, but the pursuit was unavailing, and the sea serpent when last seen was heading due east, with some appearance of haste.

Norfolk, England, 1905

This one comes from the *Evening News* (Sydney) of Monday 27 November 1905, page 3. Regrettably, the details are decidedly sketchy.

ANOTHER SEA SERPENT

The crew of a damaged vessel which was assisted into Great Yarmouth recently reported having seen a sea serpent off Hastborough Sand, a short distance from the Norfolk coast. It was sighted two miles [*3 km*] away, and the men at first thought that it was a mass of drifting wreckage. They soon became convinced, however, that the object was moving. Finally, it raised a gigantic head above the water, and dived out of sight. The serpent they declare was of immense length.

South Atlantic, 1906

This sighting had to wait 77 years to see print, and then not in a newspaper, but in a little known scientific journal. By way of explanation, cryptozoology - literally, the study of "hidden animals" - is the investigation of animals, such as sea serpents, not yet officially recognized by science. The International Society of Cryptozoology existed from 1982 to 1998, under a board of directors which included some significant professional zoologists, and published an annual peer-reviewed journal called *Cryptozoology*.

Like me, Paul LeBlond of the University of British Columbia, was a foundation member. In fact, his particular interest was the sea serpent native to his own coast, popularly known as "Caborosaurus", which we have met previously. One day he was involved in a conference at the University of Quebec, and happened to discuss his interests with Prof. Paelink of Rotterdam in the Netherlands, when the latter exclaimed: "But, my grandfather saw a sea-serpent!" His grandfather had been Captain J. Koopman who, after a lifetime at sea, returned home to write his memoirs. When he returned home, Prof. Paelink kindly sent Dr. LeBlond a copy of the relevant passage, presumably in translation.

It is now time to quote it from LeBlond's paper, the citation being:

LeBlond, Paul (1983), 'A previously unreported "sea serpent" sighting in the South Atlantic', *Cryptozoology* 2:82-84

It is a matter of frustration that we were not told the date of Captain Koopman's memoirs. Since the sighting occurred in 1906, it may have been written down twenty, or even forty years after the event. Nevertheless, his memory managed to provide a lot of detail. His vessel was steaming from the Mediterranean to Montevideo in Uruguay. He said it took place on a Sunday afternoon, which implies that he did not recall the exact date. However, he did recall that he was on watch about 3 in the afternoon, and that they sighted a sailing vessel, and that he changed course by a few degrees in order to read its name, although he could no longer remember it. More to the point, he was using his telescope when the wheelman suddenly called out to him to look to the starboard.

> I saw, about one hundred metres away, obliquely on starboard, an enormous beast whose length I approximated at about 60 metres [200 feet]. It was overtaking our ship, which appeared to be standing still, with the speed of an arrow off a bow. With the help of my telescope, I could form some idea of the monster, although only in an approximate fashion. The monstrous head and a number of enormous dorsal fins sticking out above water level, as well as its wide wake, showed the nearly horizontal posture of this giant sea-dragon or serpent.
>
> We were, at that time, about 40 sea miles (about 74 km) offshore from Pernambuco [Recife, Brazil]. The coast thus lay well below the horizon; nevertheless, I saw that town, upside down, profiled sharply against the sky: an impressive *fata-morgana*. The wheelman and I pointed out to the officers and crew the charming optical illusion. However, we chose to hide the matter of the sea-serpent, for fear of ridicule.

Koopman declared that he would never have brought the subject up again except that, the same year, two zoologists (yes, zoologists!) reported seeing a sea serpent off the coast of Brazil in 1905 only about 100 km north of his encounter. (This, of course, was the *Valhalla* sighting recorded by Heuvelmans on pp 372-3.)

As far as Koopman's animal is concerned, the impression I get is that the head, although visible, was held low down, not erect as with so many "long necked" sea serpents. However, it does seem to have been a bit on the large size. Occasionally, one comes across an estimated length of 30 metres - as long as a blue whale, but thinner - but although 60 metres is not unheard of elsewhere, it is pretty much out of the norm.

One wonders whether he was not mislead into including the wake in the length.

Kermadecs, 1906

The Kermadec Islands are situated approximately half way between New Zealand and Tonga, in the general region of 29° S, 178° E. This report comes from the *Australasian* (Melbourne) of Saturday 12 May 1906, at page 38. It is likewise very brief - something which turns out to be significant.

ALLEGED SEA SERPENT

The steamer Taviuni, which arrived at Auckland from the islands, reports having seen a "sea-serpent" about 60ft. [*18 m*] long, near the Kermadecs. Both on the outward and return voyage it approached close to the vessel, and remained visible for a considerable time. Observers describe the head as resembling that of a gurnet.

Mr. Frank T. Bullen who is sceptical as to the existence of sea-serpents, considers that the monster was a hump-backed whale.

A gurnet, gurnard, or sea robin is a fish with eyes close to the top of its head, which tapers forward like a triangle. But weren't any other details noticed, and what, in any case, is meant by "close" and "a considerable time"? Equally important, who was this Mr. Frank T. Bullen, and why was his opinion important? Well, a newspaper report of 7 May revealed that he had been lecturing on whales in Melbourne, and on 15 May 1906, his opinion on the *Taviuni* encounter was given in detail on page 3 of the *Sydney Morning Herald*.

A SEA MONSTER
MR. F. T. BULLEN SCEPTICAL

Mr. Frank T. Bullen, who has just completed his lecturing season in Melbourne, holds a brief against all sea serpents. A monster of the deep has only to lift his head above the water, and Mr. Bullen will assail with books of science in one hand and a portfolio of personal experiences in the other. A sea serpent must bring its birth certificate, its genealogical tree, and a plan of its anatomy before Mr. Bullen will grant it leave to exist. Last night, at Menzies' Hotel (says the "Argus" of May 11) he saw the telegram that a sea monster had been seen near the Kermadecs, and the sea serpent, (with its beholders) was at once under a fire of words, coming as sharp as a mate's commands.

"Who saw it? Where was it? How far away was it? Was the sea smooth or rough? Was it day? Was it night? Was the weather clear or foggy? What else did the notice about it? What made it look like a gurnet? Did they know what a gurnet was like?"

"They haven't told you much about that sea serpent," Mr. Bullen said. "Generally, a sea serpent has a beard. He raises his head out of the water, and stares at you. This fellow was a hump-backed whale. One of the most famous sea serpents that have been seen turned out to be a whale swallowing a cuttlefish. This was the old fellow himself, scooting along just at the surface, with his head out of the water. Scooting along, probably to meet his wife. So he'd be in a hurry."

"A gurnet's got big eyes, of course, and a hump-backed whale has got small ones. But the people who saw this sea-serpent saw something they thought were eyes. As for his length - well, the old fellow's wake, as he scooted along, would look like enough to a continuation of himself to a sailor. Anyhow, when a sailor sees something he doesn't understand, it's easy enough to multiply the length of it by three or four. Or by 20, for that matter. I've got a book, as thick as your leg, by Professor von Somebody, a great zoologist. He proved the existence of a sea-serpent in a dozen ways - all mutually contradictory. One of his sea-serpents turned out to be a known creature, about 15ft. [4½ m] long - but with a longer mane. Never mind the name; is got its syllables, and I can't remember them all.

"They saw this sea-serpent both going and coming to the islands. That makes it a whale story, too. 'Near the Kermadecs' is where the old fellow would live and hang about. I know the Kermadecs well; I've fished off Sunday Island, and caught all sorts of queer fish. A sailor's always finding things that he doesn't understand, and he generally puts it down to sea-serpents. I've caught flying gurnets in 14ft. of water in St. Vincent's Gulf, and I once caught a fish with purple tassels growing out of pink spots all over him. I kept him on board till everybody complained; then I threw him away. And none of the scientists at home would take my word for him. But I never saw or heard of a sea-serpent that could not be accounted for in some other way.

"There's no sea-serpent," Mr. Bullen said. "Professor Owens settled that for me. Being an Ophidian, he would have to live at the surface, and some of his remains would be washed ashore; and they never have been. A man in the United States used to show a sea-serpent in a museum, till somebody found out he had dug up the pieces of it in Nebraska or somewhere, and 'faked' it. It was a mixture of a Plesiosaurus and a Pterodactyl, or something like that - a pre-historic land animal.

"Take that sea-serpent away," Mr Bullen concluded, handing back the telegram. "He's the least healthy I've met for a long time."

Personally, I consider the criticism quite valid, particularly the second paragraph. All of that information is essential to make a proper determination of the matter. In point of fact, in dealing with Australian cases, I have come across a number where a whale was mistaken for a sea serpent. In fact, the "sea serpent" off Montague Island, NSW was so well described it was easily recognizable as a humpback whale. Why Mr Bullen was so convinced the Kermadecs creature was a humpback and not some other species of whale is not clear. The details are simply inconclusive. Indeed, it is still theoretically possible that it really was a sea serpent, but the information is just not sufficient.

Scotland, 1906

This event in fact took place in late September 1906, but it seems to have floated around in the ether for three and a half months before being printed in an Australian newspaper - in this case, the *Chronicle* of Adelaide, on Saturday 12 January 1907, at page 27. As was usual in such cases, it was simply reprinted from whatever original journal they took it, without any attempt at explanation. In those days, of course, there was no internet, so probably none of its South Australian readers had any idea where Dunottar Castle was. It is, in fact, a ruined fortification overlooking the North Sea half way up the east coast of Scotland, at approximately 57° N, 2° W. You will note that, although the witnesses were using field glasses, the "monster" was quite a distance away, so you may draw your own conclusions.

SEA SERPENT'S RETURN

A sea monster was observed off Stonehaven coast, near Dunottar Castle, on September 29. A party of Volunteers were having a shooting match, when one of their number observed a commotion in the otherwise calm sea. A huge body appeared above the surface, and as it was less than a mile [*1.6 km*] off, the Volunteers were able, by means of their field glasses, to make out that it was some sort of sea monster. As far as could be judged, its length was 50 yards *45 metres*]. The body was narrow and of a dark color, and was surmounted by a number of short fins, with protuberances, probably two or three on each side of the main fin. The monster was going northward at the rate of six miles an hour [*9½ kph*]. It remained above water for half a minute at a time. It did not blow as it came to the surface, and those who saw it are quite convinced it was not a shoal of porpoises. The monster was sighted for half an hour and observed at half-past 10 on the same night about two hundred yards off Stonehaven beach by Sergeant Wright and Constable George. It was clear moonlight. The monster encountered a ledge of rocks called the Brachans and rose high in the air, but finding its way impaired fell back into the water. The police compared it to a large trawler getting on the rocks, and thinking it was coming ashore they were prepared to watch its movements from behind a boat.

North of Borneo, 1907

This story was two months old by the time it was picked up by an Australian newspaper - in this case, on page 4 of the *Newcastle Morning Herald and Miners' Advocate* of Thursday 18 July 1907. It is interesting that it took place not far from the city of Labuan, which also gave its name to the ship involved. Note that it was apparently recorded in the ship's log.

A SEA SERPENT

The "North Borneo Herald" states that at noon on May 14, lat. 7.5 N., long. 117.5 E, a sea serpent was seen from the deck of the steamer Labuan by the engineer in charge, a passenger, and a native serang. It appeared to be "at least 50ft [*15 metres*] long, and moved in a wriggling motion on the surface of the water" in an opposite direction to the ship about 200ft [*60 m*] distant. It was visible for about four minutes, concludes the log entry, which was signed by the three witnesses of the gruesome spectacle.

South Atlantic, 1909

This report was originally published in the *Boston Herald*, on who knows what date, and was taken up by the *Newcastle Morning Herald and Miners' Advocate* (NSW) on Tuesday 21 September 1909, on page 6.

As is often the case, the reader is simply expected to know the places referred to. Thus, the event took place on a voyage from Penarth to Santos. Since the ship was British, and the name sounds Welsh, I presume the city of origin was the original Penarth in Wales, rather than its namesake in Delaware, USA. Likewise, by Santos is probably meant the big port in the city of São Paulo, Brazil. However, the story didn't come out until the ship docked at Boston, which implies a three cornered voyage. The huge size of the creature, and its description as like a giant lizard with huge horny scales, appear to be unique - which makes me doubt its veracity.

SEA SERPENT RACED SHIP

Entered in the permanent log of the British steamship Mereddlo, Captain Clark, is a record of a sea monster sighted while the ship was on a passage between Penarth and Santos. Chief Officer Neal S. Murray was in charge of the bridge at the time, and a Greek quartermaster was at the wheel. The quartermaster, who first sighted the monster, was almost petrified with fear, and was at the point of permitting the big freighter to take her own course.

"It was like this," explained the chief officer when the Mereddlo docked in East Boston. "The ship was 500 miles [*800 km*] from Santos. I saw the Greek acting strangely, and followed the direction he was looking.

"My hair nearly stood on end at the sight. Swimming parallel to the ship was a monster lizard. It was as big as a whale. The ocean fairly seethed as it propelled itself with enormous dragon's claws. A head as big as a pilot house and one coil of the beast's neck were above water.

"For a distance of nearly 300 feet [*90 metres*] the sea heaved and was lashed into foam. I think the lizard was fully the Mereddlo's length, and I feared for the safety of the steamship, as the creature, mailed in huge horny scales of a dark green colour, swerved as if to come alongside. It had a saw-like ridge on its back, and its girth was fully as great as that of a whale.

"After the serpent had raced the ship several minutes it humped its back and sounded. The swash was from its commotion shook the ship and sent spray over the starboard rail.

"I have followed the sea many years and, mind you, I m not given to fancies. That creature so impressed me that I entered he incident in the gerap [?] log, and later made a permanent record of it."

The Mereddlo's crew substantiate Officer Murray, while the Greek quartermaster admits he did not recover from the shock for several days. - "Boston Herald"

Off West Africa, 1911

This incident was published in *The Times* 22 years after the event, but was apparently written immediately after the event. I have taken the story from page 2 of the *Kalgoorlie Miner*, Thursday 2 February 1933.

THE SEA SERPENT
A CAPTAIN'S EXPERIENCE

In a letter to 'The Times' Sir William Brandford Griffith writes : — In a leading article in your issue of October 30 you referred to the reticence of sailors on the subject of the sea serpent and the invariable ridicule the subject excites. I enclose a note made immediately after the occurrence of a well authenticated case. I was on deck at the time, and was called, but half asleep in a deck chair, I did not realise what was passing until too late. Mr. Punch, who made the note, is a trained botanist and a reliable observer; he tells me that he sent the note to Professor Ray Lankester, who treated it with scorn. When I asked the skipper if he proposed to enter the occurrence in the log the reply was, 'If I did I should lose my ship.' That explains a captain's reticence.

S.S.Aro.

On Saturday, March 11, 1911, the s.s.Aro was proceeding homeward bound from Sierra Leone towards Las Palmas. At about 11.45 a.m. her position was about lat. 18.15 N. long. 17.34 W. Captain Pooley was on deck and observed on the starboard bow a mass in the water, which at first he took to be seaweed. On approaching nearby he observed that it was moving, and then distinctly saw that it was a living creature. He called to a passenger, Mr. Punch, who was on deck, and directed his attention to the creature in the water, saying that it appeared to be a sea serpent. At the same time he saw the creature raise its head above the water. The head was shaped like that of a turtle, and immediately behind the head were a pair of diamond-shaped fins. The head, and fins were black or dark in colour. The creature lowered its head into the water, exposing a section of its body above the water. The section was round in shape, about 18 in. to 24 in. [*45 to 60 cm*] in diameter, and of brownish-white colour on the side exposed to view. At some distance behind and submerged another light coloured section of the body could be seen. Captain Pooley estimated that the length of the creature would be not less than 40 ft. [*12 metres*] Its movements appeared to be sluggish. The steamer passed within 30 ft. to 40 ft. [*9 to 12 metres*] from it, and there was time to observe it carefully.

Mr. Punch, on hearing the call from Captain Pooley, saw plainly the two light-coloured masses in the water in front as the steamer approached. He did not see the head, but distinctly and plainly saw one section of the body about 10 ft. to 12 ft. [*3 - 3.6 m*] in length exposed above the water, with the other light-coloured section some distance behind. He called to another passenger, Captain Craven, to come and observe the creature. Captain Craven came and distinctly saw the two light coloured masses in the water as the steamer passed at a distance of about 20 ft. to 30 ft. [*6 to 9 metres*]. There was a strong head wind, and the creature passed astern somewhat slowing, so that there was time for careful observation. Messrs, Punch and Craven are clear that the body was round in shape and of about 18 in. to 24 in. in diameter. Captain Craven agrees with Captain Pooley that the length of the creature would not be less than 40 ft. The water was quite clear, and the body could be plainly seen under the water.

I hereby certify to the foregoing statement, and subscribe my signature.— Ric. Pooley, master, R.M.S. Aro.

The above statement is correct. — Cyril Punch, district commissioner, S. Nigeria; J. Craven, captain,West A.F.F.-, S. Leone.

Tasman Sea, 1911

This time we have another appearance in the waters between Australia and New Zealand. The report comes from the *Age* (Melbourne) of Friday 22 December 1911, on page 7.

> THE SEA SERPENT AGAIN!
> "SIGHTED" IN THE TASMAN SEA
>
> SYDNEY, Thursday.
>
> The sea serpent has turned up again - this time in the Tasman Sea. Officers of the cargo steamer Strathardle, which arrived

to-day from New Zealand, report having seen "the monster". It appeared about a quarter of a mile [*400 metres*] from the ship on Sunday afternoon last. The lookout on the bridge suddenly reported a strange looking object ahead, and all eyes were turned in the direction. The wind was blowing strong from the westward at the time, and the steamer, in consequence, could not get very close to it. Examination was made with the aid of glasses. One of the officers described it as being some 200 feet [*60 metres*] in length, with a head like that of a crocodile. A large fin showed out of the water, and the body, which appeared to be of bright colors, tapered away considerably at the tail. When seen the serpent was travelling very slowly, and it was soon lost to sight as the steamer forged ahead. The position of the Strathardle at the time "the monster" was passed was about 200 miles [*320 km*] off the New Zealand coast.

Chapter 7
1914 and After

The First World War did not see the end of sea serpent sightings, though I have noticed that reports declined dramatically after the Second World War, at least in Australian newspapers. It was as if that major conflict kept lesser matters out of public view, and when it was over, people had forgotten that it was once respectable to see sea serpents and to report them in the press. Be that as it may, the previous three decades was still rife with them.

Off Borneo, 1914

Here we have what, in the Australian vernacular, would be called a "humdinger". It's a highly dramatic account of what appears to have been a hexed voyage. The highlight, sandwiched between a series of fatal accidents, was an encounter with a sea serpent which appeared to want to eat a fallen sailor. Since such behaviour has been so rarely reported, the truthfulness of the story must rest in doubt. (Then again, perhaps the rarity of such reports might be simply explained by the extreme rarity of seamen falling overboard when a sea serpent is in the vicinity.) In any case, the story is too good to pass up. It comes from page 2 of the *Express and Telegraph* (Adelaide) of Saturday 8 August 1914.

WITH THE GHOST AT SEA.
THE SERPENT TERROR.

More like chapter from a sensational novel than an incident in real life are the adventures of of the British tramp steamer Strathspey which arrived in New York on July 1, after a remarkable voyage of one hundred and fifty days from Glasgow to the Far East, via the Cape of Good Hope, and back to New York by way of the Suez Canal. Of her complement of 38 officers and men only eight are whites, the remainder being Chinese, Arabs, and Lascars, says the "Central News" in an account of the voyage. Off Port Natal a Chinese stoker was killed by falling into the hold, and for three days afterwards the steamer merely drifted, the other stokers refusing to work on the ground that the ghost of the dead man was prowling about the stokehold. In the Canton River one of the Chinese coolies working the cargo was knocked on the head by a heavy chain and instantly killed. Off Malta, the chief engineer,

James McMurray, jumped overboard, and nothing was seen of him again, although a prolonged search was made.

But the most remarkable adventure of the voyage occurred off the coast of Borneo. According to statements made to the New York newspaper reporters, Mohammed Singh, an Arab sailor, fell overboard from one of the boats he was cleaning. Singh, a powerful swimmer, was nearing the lifeboat when a commotion arose in his wake, and the crew of the boat saw a great green sea serpent raise its head several feet above the waves as if about to seize the Arab sailor in its capacious maw. Singh head the noise and felt the hot blast from the monster's lungs on the back of his bronze neck. He turned half round and then, with a cry of "Allah Kerim," he made a tremendous effort to reach the boat before the sea serpent could seize him.

The Arab sailors on board the boat bent their backs double on the oars and gave a mighty pull, which enabled Singh to be hauled aboard breathless just as the serpent opened its mouth to grab him. Finding he was out of reach, the monster bit the rudder off the boat in its rage. Chowder Loll, who was steering at the time, fell in a faint from fright. The boat was steered back alongside the steamer with one of the oars over the stern. The sea-serpent evidently had been scared off, for he was seen in the distance steering due east at the rate of 30 knots an hour [*55 kph*].

After the Strathspey left Port Said on June 18 for New York it was noticed that James McMurray, who was over 60 years old, was very melancholy and walked about the deck a good deal. He was very fond of the parrot, and after talking to him one morning when the steamer was off Malta, he fell overboard. The parrot, Toko, shrieked "eight bells" until the chief officer heard him and saw the chief engineer's coat and vest and cap by the rail. Then he realised what had happened. Captain Jones had the ship stopped and went back 14 miles, but could not see anything of the old man. The Strathspey is a steel screw steamer of 4,432 tons, built by the Grangemouth and Greenock Dockyard Company in 1906, and owned by the Strathspey S.S. Co. Ltd. (Messrs. Burrell & Son).

A couple of trivial points. "Allah Kerim" is Arabic for "God is noble/generous", but Singh in an Indian name. I therefore conclude that, in this account, "Arab" meant "Muslim". As far the sea serpent goes, it is a pity that the journalist did not interview anyone with a close up view of the animal in order to obtain more details, other than the fact that it was green (a very unusual colour) and that it had a "capacious maw". On the other hand, back in 1965 a skin diver described how, off the coast of Florida in 1962, his four companions were apparently killed and eaten by some sort of sea monster. The story was reprinted on pages of 524 - 5 of Heuvelmans' book. Although the tale sounded fantastic at the time, it does tend to corroborate the present account. It just goes to show that you should never discard any report, no matter how unlikely it appears.

U28/Iberian Incident of 1915.

This is a "classic" sighting which has been retold many times. Essentially, in 1915, the British steamer, *Iberian* was sunk by the German submarine, U28. In 1933, when the Loch Ness Monster was in the news, the U-boat's captain, Baron von Forstner narrated in a German newspaper how, immediately after the sinking, an underwater explosion threw to the surface a lot of wreckage, plus a mysterious creature which resembled a crocodile 20 metres in length.

None of the survivors of the *Iberian* noticed such an event, and considering the lapse of time, there was always suspicions about the veracity of the account. It is germane, therefore, that *The Western Australian* (Perth) published, on page 6 of its issue of Saturday 28 September 1935, the following letter from a reader, Matthew A. Utting:

A SEA-SERPENT HOAX.
To the Editor.

Sir, — In Life and Letters on Saturday last appeared the review of a book, "Mysteries of the Great War," by Harata T. Wilkins, wherein the author mentions the appearance of a sea-serpent, as seen by the commander of a German submarine which torpedoed the ship Iberian. I think I can give a different view of the matter. I was a passenger on the Centaur some weeks ago, travelling north from Fremantle, and my cabin-mate was a German traveller going to Batavia [*now Jakarta*], a Mr. Wegener. During our conversations he told me the story of the sea-serpent. Wegener has a close friend, who during the war was the commander of a German submarine. This commander, while reading about the monster of Loch Ness recently, decided to play a joke on the German public. He wrote to the newspapers stating that during the war, just after the Iberian sank, she exploded under the water, and a sea-serpent was blown into the air. He wrote the description of this animal on a sheet of paper, which he placed in the log-book. As all the log-books are kept, a large number of people, on reading about the sea-serpent rushed to investigate. They found the log-book page containing the report of the sinking of the Iberian, but no sheet of paper describing a sea-serpent. When they questioned the commander he said, "Well, it must have become lost," so that the public were nicely hoaxed. Evidently the author of the book has also been fooled, as he includes the story without any reservations. This story is vouched for by Mr. Wegener, and I feel sure that he would verify my statement. The submarine mentioned was the first one made by Germany. It was captured later by the Allies, but was returned to Germany after the war, and is now in a museum. [*In point of fact, it was sunk.*]

— Yours, etc.,

MATTHEW A. UTTING.

Cottesloe

Now, before we get carried away, remember that skepticism cuts both ways. Just as we should hesitate to accept a fantastic story at face value, we should not leap uncritically at any third hand tale which appears to explain it away. Nevertheless, I consider this one more large nail in the coffin of the *U28/Iberian* sea serpent account. Who would have thought that a tale in a German newspaper could be debunked by an obscure newspaper on the other side of the world?

Baltic Sea, 1917

Here we have an example of the randomness in the way these stories are published. In this case, it was picked up by two obscure Australian country newspapers months apart. The first one was a short paragraph on page 35 of the *Albury Banner and Wodonga Express* (NSW) of Friday 23 February 1917. It refers to a Swedish officer having lately written to the press. However, a more detailed citation of the officer's account was published in the *Cobram Courier* (Vic) of Thursday 3 May 1917, on page 7, so I shall provide this version. Here the officer is said to have written to *Nature*, which is a very prominent scientific journal, though I am unaware of which issue carried it.

SEA SERPENT SEEN IN THE SEA.

Major O. Smith, an officer of the Swedish army, has described in "Nature" a sea serpent which he saw in the Baltic Sea, near Stockholm. "At 2.23 p.m." he writes, "we suddenly saw a movement of the water like the ripple of a wave less than 300 feet [*90 metres*] from us. The sea was calm and there was no boat or anything else that could cause such a movement. Looking more attentively, all of us saw very distinctly a head like that of an enormous serpent, larger than that of a man, slightly elongated, surmounting a serpentine body about seventy-five feet [*23 m*] long. The creature was undulating, making at least ten distinct curves, and a large part of its posterior region was above water. We watched this strange creature for more than a minute swimming at a speed of about two knots [*3½ kph*]. I have seen many dolphins and whales and I know their movements. Those of this sea serpent were very different."

Atlantic side of South Africa, 1918

This one comes from page 6 of the *Darling Downs Gazette* (Qld) of Monday 21 October 1918.

SEA SERPENT AGAIN
A STORY FROM AFRICA

That old dear, the sea-serpent, has appeared again. Ethelbert G. Fotheringay says so, and as he is not a German his word is entitled to respectful consideration. In fact, he was loathe to tell the story, for he is hep to the merry ha-ha that usually greets sea-serpent tales. Of course, this is the open season, and they may be caught at any time now off seacoast summer resorts. But this serpent chose the coast of Africa for his appearance - probably he was disturbed in his deep-sea lair by a prowling submarine, and took it on the run for the south.

Mr. Fotheringay has been in Africa for two years gathering rhinoceros hides and ivory for a Chicago firm. He saw the serpent three months ago while on the way from Swakopmund, formerly German South-west Africa, to Capetown, and this is the way he told the story to a New York 'Times' reporter, reluctantly as has been said.

'I was on board the old African steamship Lum-Lum, which carried a Chinese crew with Dutch officer, and commanded by Captain Johann Van den Woof, one of the oldest skippers on the coast, a lifelong teetotaler, and a Baptist. There was only one other white passenger besides myself, Guy de Jolipas, the famous French gorilla-hunter, and about two hundred Hottentots and Kaffirs.

HEAD LIKE A PORK BARREL

'It was a sweltering afternoon and the ship was about 150 miles [*240 km*] north west of Capetown. The temperature was 105 [*40.5° C*] in the shade, with a copper-coloured sky and the sea like boiled oil. Guy, the gorilla-hunter, had just thrown a chatty at the head of Oolu, the Hottentot cabin-boy, because he had brought him a bottle of beer without ice, when I heard a wild yell from deck and saw the panic-stricken natives trying to get down the after-hatch looking over the port side. I saw the weirdest monster that one could possibly imagine, afloat or ashore. When I tell you calmly that the head of the animal, which I realised at once was the sea-serpent, was as large as a good-sided pork-barrel, I do not exaggerate. I refer to the ordinary 500 [*unclear*] pound barrel and not the [?] tierce of beef which is usually 350 pounds [*159 kg*] or more.

'The sea-serpent's head was about eight feet [*2½ metres*] above the surface of the sea, and about three feet [*90 cm*] across in the widest part. Its face was covered with bristly spikes which stuck out at angles, and the large, round eyes gazed curiously at the steamship in a reproachful manner, as if the noise of the propeller had disturbed its afternoon siesta.

'The neck was no more than twelve inches [*30 cm*] in diameter, and was partly hidden by dark, hard-looking barnacles. I could not say exactly how long the sea-serpent was, but judging by the last ripple when it moved I think 150 feet [*45 m*] would be about the mark. Captain Van den Woof was very much excited as he stood with his big telescope on the bridge examining the marine monster. 'Gott fur dicker,' he shouted, 'this was the big sea-serpent the old Danish skipper Jensen reported three months ago at Capetown, and the people said he was crazy,'

SHIP DOES FIVE LAPS

'The captain gave orders to the officer on watch to steam around the sea-serpent carefully and get as close as the ship could go without rushing into needless danger. Five times the Lum-Lum circumnavigated the sea-monster, which turned its massive head slowly, and regarded the vessel with a wistful look as if he wanted to speak to us and tell about its travels around the world. No-one had a camera on board, and the finest chance to snap the sea-serpent was lost. Guy, the hunter, had one when we left

Swakopmund, but he broke it on Oolu's head two hours later and threw the debris over the side. He fired his express rifle at the monster several times, and the skipper peppered away from the bridge with an old Snider rifle, but the bullets glanced off its hide without having any perceptible effect.

'Finally the captain gave orders to resume the course, and the Lum-Lum steamed away for Capetown. The last we saw of the sea-serpent astern was the great barrel-shaped head wagging slowly up and down, followed by a big commotion in the water, and then he disappeared beneath the surface. Judging by the course taken, the serpent was going an easy thirty-knot [55 *kph*] gait towards the Bight of Benin.'

Fiji, 1923

Once more, we are back to Fiji. To be fair, it took only a couple of weeks for this story to reach Australia, where it turned up in a couple of minor rural newspapers, but, for reasons explained in the second last paragraph, it took four months for the story to break. In other words, although it took place in 1923, it was not reported until 1924. This is taken from the *Advocate* (Burnie, Tas.) of Friday 8 February 1924, on page 5.

SEA SERPENT SEEN.
Mission Station Experience.
HEAD LIKE A HORSE.

SUVA, January 25 - A remarkable story of a huge sea serpent having been seen comes from the Methodist Station on the island of Taveuni. Nurse Davis, who was on a professional visit to the Mission Stations, was looking out over the Some Somo Straits, which lie between the island of Taveuni and that of Vanua Levu, when she saw a huge head, which she describes as being something like that of a horse, rear itself out of the water. A thick body of a dull brown color, followed, until the head was reared above the waves to a height of 30 feet [9 *metres*], suggesting the huge length of the monster beneath the surface. Nurse Davis called out in horror, and her cries brought the Rev. Lelean to the scene. Through his field glass he had an excellent view of the sea serpent. For some time the head and neck of the thing swayed slowly in the air. Then it slid silently beneath the waves and disappeared. This happened on October 9, but shortly afterwards her professional work took Nurse Davis to the Lau Group, and she only recently returned to Suva with her story. The description of the sea serpent is very like that of a monster alleged to have been seen near Noumea by a sailing vessel recently.

A neck 30 feet long is not unknown (they were reported, for example, off South Africa in 1901 and off New Zealand in 1932), but even if the length were overestimated, and it were only 20 feet, the implications are significant. As I have pointed out before, mammals - even giraffes - possess only seven neck vertebrae. Therefore, if the animal were a mammal, its neck would be too stiff to manoeuvre underwater.

As for the Noumean monster, this also appears to be one which has slipped through the researchers' net. I have no record of it. The same goes for Captain Jensen's sea serpent, mentioned in the previous article.

Java Sea, ? before 1924

Here is something which can only be described as an orphaned article. It just happened to omit some cardinal features, such as the name of the witness, and the year. Although it was published in 1924, the reference to sailing vessels, and the phrase, "in those days" suggest it might have happened some time in the more or less distant part. I have no idea where the port of Angier Road might be, but I presume it is/was somewhere in the home countries, and nowhere near the site of the encounter: the Java Sea, between Java and Borneo. Anyway, for what it is worth, here is the story, which appeared on the front page of the *Bathurst Times* (NSW) of Tuesday 4 November 1924.

Another Sea-Serpent

Some people still see sea-serpents. The appearance of such monsters used to be put down to the brand of whisky favored by the ship's passengers. But a seasoned tourist vouches for the following: - "In crossing the Javanese Sea, on the way to Angier Road, the great port of call for homeward-bound sailing vessels in those days, we saw, one forenoon, an enormous water serpent passing over the channel between two islands, just as we were emerging from it. It must have been nearly fifty feet [*15 metres*] long and as thick in width as the largest of our coir hawsers. It was holding its head some two feet [*60 cm*] or so above the water, and moving at a rapid rate."

British Columbia, Canada, 1926

Here we have another of those huge 30-foot necks sticking out of the water. This time it was off British Columbia which, as we know, is the haunt of the famous "Cadborosaurus", which is usually described as short-necked, humped, and undulating. This caused Dr Heuvelmans to state in a footnote on page 473 of *In the Wake of the Sea-Serpents*:

> [I]t was certainly not Caddy that Captain House of the Canadian Government Fishery Patrol saw looking "like a 30 ft. telephone pole" near Hecate Strait between Queen Charlotte Islands and the mainland.

Obviously, Heuvelmans was citing a brief reference in some secondary source. Well, here is the full story. I have taken it from the *Newcastle Morning Herald and Miners Advocate* (NSW) of Saturday 29 May 1926, on page 7. One only wishes that the three sketches provided by Captain House had been included. As far as the location goes, it was not, strictly speaking, in the wide Hecate Strait. Rather, it was in a very narrow passage between a group of small islands and the mainland. The Wikipedia gives the co-ordinates of Wright Sound as 53° 20' N, 129° 14' W.

SEA SERPENT AGAIN

Seafaring men are already identifying the sea serpent reported to have been seen off the coast of British Columbia a fortnight ago with the one seen in the same vicinity eighteen months ago, says the Vancouver correspondent of the N.Z. "Herald," writing on April 7.

Captain House, of the Canadian Government fishing patrol, was on his way north to Prince Rupert, and had reached a point near Hecate Straits, which separate Queen Charlotte Islands from the mainland, when he saw the monster. He is an officer in whom his fellows place high trust. They say, if Captain House said he saw a sea serpent, he saw one; that is enough for them. He had plenty of time to observe the sea serpent, and made a drawing of it, in three positions - as it emerged from the water, when it was most out of the water, and when it was slipping back into the depths.

The following signed statement from Captain House has appeared in the Vancouver Province newspaper: - "I have prepared three sketches of the sea serpent sighted off Cape Bridge, opposite Wight Sound, at 2.45 p.m. on March 16 [?], coming towards the south end of Grenville Channel. The head was about 18in. [*45 cm*] wide and possibly 2ft 6in [*75 cm*] long. The thing remained erect for about half a minute, and then disappeared spirally, as it had come. When submerged, it churned up the water, and left a wake for a long time, like a school of porpoises, moving outwards toward the sea, whence it had come."

Captain House remarked that the sun was shining from the clouds at the time, and gave the monster a greenish-gold appearance. He said he was familiar with most sea creatures, and was positive it was nothing he had seen before. It had the appearance of a telephone pole, as it raised its head 30ft. [*9 metres*] above the water.

The capture, last week, of a serpent-like fish at Powell River, lends colour to the belief that it is the young of Captain House's sea serpent. The Powell River Company have put it on ice, and are sending photographs to the Fisheries Department. It is five feet [*1½ metres*] long, with a head like that of a wolf, about the same size as an English pug-dog. The end of the body tapers to a point and skin-like fins at the side extend the entire length.

Los Angeles, ? late 1926

One imagines the editor in Perth, Western Australia saying, "Where in the U.S.A. is this place, Venice? Oh well, it's a good story. It'll fill up a paragraph, so we might as well run it." But none of the readers were left with any idea of the geographic context in those pre-internet days. Venice was a Californian seaside town which was annexed by Los Angeles in 1926. It has a popular amusement pier, and during the 1920s its mayor was Clinton Gordon Parkhurst, who now has a building named after him in Santa Monica. Presumably, local newspapers provided more details concerning who, when, and what than is provided in this short paragraph which, if accurate, indicates something rather strange. It comes from the front page of the *Daily News* (Perth) on Monday 3 January 1927. This suggests to me that the events took place in late December 1926.

AMERICAN SEA SERPENT

A giant "sea beast" sighted off shore between Liek and Venice piers (U.S.A.) caused a buzz of excitement among beach residents. A permit to kill the huge fish was obtained from the Venice police by residents along the waterfront. The "beast," according to Mayor Parkhurst, was about 35 feet [*10½ metres*] long and had six fins projecting out of the water. Fishermen who sighted the strange water creature believe it was attracted by the presence of an unusually large number of dead seals in this vicinity.

Gibsons, British Columbia, 1927

The city of Vancouver, of course, is situated on the mainland of British Columbia, Canada. Just to the northwest lies a triangular stretch of water known as Howe Sound. Gibson lies just on the other side of the sound, at 49° 24' N, 123° 30' W. This was the location of the following series of encounters - in a stretch of water known as the Strait of Georgia, separating Vancouver Island from the mainland. Of course, as readers should now be aware that this whole area is a known haunt of "Cadborosaurus". The story is taken from the *Mercury* (Hobart) of Friday 22 July 1927, at page 12. Needless to say, the vertical undulations reported clearly reveal that the animals are mammals.

SEA SERPENTS
SEEN BY FISHERMEN.
AMAZING STORY FROM VANCOUVER.

Two sea serpents measuring more than thirty feet [*9 metres*], and carrying their heads nearly six feet [*1.8 m*] above the sea level, are reported to have been seen by three Vancouver men, and many residents of Gibson's Landing, British Columbia (says a message from Vancouver to the New York "Herald-Tribune").

ENORMOUS HEADS.

The heads of the creatures were said to be about two feet [*60 cm*] broad by one foot high, with enormous mouth and bulging eyes. The neck tapers slightly from the body, which is thirty inches [*75 cm*] thick. Coils, twenty feet [*6 m*] behind the head, were two feet in diameter, and loose skin hung below the lower jaw, probably a pouch. The head resembled a snake's, but the serpents swam by an up and down wave-length motion. They were devoid of anything resembling fins or scales, the skin being pale pink, blotched with dull yellow.

This description is given by Frederick Parnell and William Park, of Vancouver. Parnell was fishing near shore at Gibson's Landing, when two serpents broke water within 25 feet [*7½ m*] from his boat, reared their heads nearly six feet, and blew heavily, but did not eject water like a whale. Terrified, he remained silent, watching. Moving their heads from side to side like snakes, the creatures moved rapidly away coils appearing and disappearing behind the heads, which had been lowered slightly.

DIVED FROM SIGHT

Recovering his nerve, Fred, shouted to his brother George, on shore, "Look, quick." Immediately he did so both serpents lowered their heads into the water and dived from sight.

William Park was with D. Smith in a row boat when they saw a serpent near the same spot. It appeared fifty yards away, and blew heavily, its head nearly five feet [*1½ m*] above water. Both thought it was a seal or sea lion, but on observing it closely they saw it was neither. After blowing for half a minute the creature dipped its head. As it dived several coils broke water before the tail disappeared.

Fishermen who have also seen the monsters are worried, as fish have almost disappeared from the bay. Plans are being discussed for an organised effort to capture the creatures.

W. Messenger Racher, of Gibson's Landing, has signed a statement declaring he saw the sea serpent twice within 300 yards from the shore off Gower Polat, on April 18. The creatures appeared within 40 feet [*12 m*] of his boat.

Jersey, English Channel, 1928

Here we have a brief, but graphic account from the *Advertiser* (Adelaide), on Friday 6 July 1928, page 15.

THE "SEA SERPENT" AGAIN.

Pilots on duty near Corbiere Lighthouse, Jersey, are confident that they have seen a sea serpent. A huge head with bulging eyes appeared a few hundred yards from their boat, they reported in May. "Its head was held upright for some seconds, and, snorting loudly, it dived back and swam off at great speed in a southerly direction," they said. In all their experience, the pilots stated, they had never seen such a large or repulsive-looking fish. It appeared to have horns on its head and spikes or spines from the head downwards.

Off West Africa, 1929

Originally published in a prestigious English newspaper, this report was taken up by one in rural Queensland, the *Western Star and Roma Advertiser* (Toowoomba) on Saturday 26 September 1931, on page 5. Note that the location was not far from that of the S.S. *Aro* sea serpent of 1911, mentioned earlier.

ANOTHER SEA SERPENT

Mr. J. J. R. Smythe, of Notinghill, London, in a letter to "The Times Weekly Edition," makes an interesting contribution to the immemorial controversy concerning the sea serpent. "While travelling to England from Durban in Grantully Castle on March 1, 1929," he writes, "I was on the promenade deck with my wife and my son, who was fourth officer of the ship. At 7.15 a.m., the ship being then in latitude 15.38deg. N., longitude 17.39deg. W., I noticed a big disturbance in the sea ahead. Looking through my binoculars I saw that it was caused by a swimming reptile of some sort. I handed my glasses to my son and asked what he made of it. He said that it was a sea serpent, and that he had seen one on a former voyage; but with usual reticence of those who make their living on the sea he had said nothing about it. The creature rapidly approached the ship and passed us going south at a distance of about 600ft. [*180 metres*]. From the time I first sighted the serpent till it passed out of sight astern could not have been more than three minutes. I estimated its length to be 100ft. [*30 metres*] and its diameter to be 4ft. [*1.2 m*]. Its colour was a dirty yellowish green with large white spots on the body. I saw three distinct undulations or curves above, though not clear of the sea; my son saw the head and said that it looked like that of a snake."

Aden, 1933. On page 432, Heuvelmans has a footnote:

In 1935 a dozen people again saw one [ie a sea serpent] in Largs Bay, also in the Gulf of Aden.

Wrong on nearly all accounts. The event took place in 1933, not 1935. It did happen near Aden, but Larg's Bay (does it sound like an Arabic name?) is not there, but is next to Adelaide, South Australia. More to the point, it was the name of a ship which docked at Southampton on approximately 22 May 1933, and presumably immediately told her story to the English press. This was subsequently picked up by a large number of Australian regional newspapers, with minimal verbal variation, over the following three months - proving that old news is not necessarily bad news. I shall take this one from the *Cairns Post* (Queensland) of Wednesday 24 May 1933, page 9.

'SEA SERPENT'
SIGHTED OFF ADEN
LARGS BAY STORY
(Australian Cable Service)

LONDON, May 23.
The s.s. Largs Bay, which has arrived at Southampton from Australia reports having sighted a sea serpent off Aden. It jumped partly out of the water about a cable's length from the ship. Eye-witnesses say that they saw about 20 feet [6 m] of the monster. It appeared like an elongated fish and reared itself twice from the sea. It had a bulging head, the circumference of which was greater than the body. A long spike or tongue projected from the snout.

England, 1937.

This was originally published in an English journal, the *Daily Sketch*, which I am unable to access. I shall therefore quote the first of the five Australian newspapers which picked it up over a period of two weeks. It was on page 8 of the *Advocate* of Burnie, Tasmania in the issue of Thursday 30 December 1937,

SEA SERPENT STORY FROM ENGLAND

A white sea serpent, humps and a face like a camel and eyes bigger than golf balls, has been seen by Mr. Harold Groves, head Gardener to Cyril Maude, the actor. Mr. Groves is a steady man, says Mr. Maude, who accepts as fact what the gardener saw as he was fishing one evening in Redlap Cove, Dartmouth (Eng.).

"I was looking out to sea," Groves said, "when I saw, about 50 yards away, a creature swimming in the water. "It had three humps, and there was at least 12 feet [3.6 m] of its body above the surface. I picked up my fishing gear to climb to the cliff-top for a better view, but when I raised my head again I was startled to see, right in front of me, the head of the monster. It was less than five yards away, and it stared at me for about 10 seconds before it submerged.

Tuft On The Top.

"The monster had a face like a camel. The head was about two feet [*60 cm*] long, and a tuft of hair on the top of the skull was quite thick. Otherwise the head was entirely hairless. The skin was almost white. It had large, unblinking eyes, bigger than golf balls, and it was very uncanny the way it stared at me. I have never credited stories of sea monsters, but there was no mistake about this one."

Mr. Maude did not see the monster, but his comment to me was: "My gardener is a very steady and respectable fellow who has had a lot of experience of the sea. I am certain he did see the thing he has described to you."
-"Daily Sketch."

It would have been nice to have been told the time of the sighting, or at least the lighting conditions. At Dartmouth in the last week of the year, sunset is about a quarter past four, and dusk just before five. However, it is unlikely he was fishing in total darkness, and the short distance lends credence to the sighting. Apart from the unusual paleness of the animal, which might have been a lighting effect, the description fits well with the "long necked" sea serpent, and it is unlikely Mr. Groves would have been familiar with that.

South Africa, 1939

This article is from the *Sunday Times* (Perth) of 30 July 1939, on page 9. It sounds like a pretty standard "long necked" sea serpent. Unfortunately, I have no information about the alleged sighting five years before.

SEA SERPENT
(By Air Mail)

Considerable excitement was prevalent at Umtentweni recently when a large sea serpent was seen by a European and several natives a few hundred yards out to sea. The serpent was moving towards the north. It was seen clearly.

At times it reared its head about 10 feet [*3 metres*] out of the water and then dived. Portions of its massive body were visible. The body appeared to be of a girth equal to that of a bullock. Its length was estimated at more than 100 feet [*30 metres*].

About five years ago two residents walking along the Umtentweni beach at dusk came upon a sea serpent in one of the large pools near the river mouth. In this case they were within a few feet of it. At the time this, discovery attracted a great deal of attention, especially as a few days later several other people saw the monster, which was estimated to be about 80 feet [*24 metres*] in length.

Devon, England, 1987.
It is a sad fact that, after the Second World War, sea serpent reports became extremely rare. Heuvelmans had been able to collect a reasonable number up to the 1960s, but those he missed have not turned up in Australian newspapers. I find it hard to believe that all these unusual animals suddenly became extinct. More probably, after that terrible struggle had distracted people's attention from lesser matters, people forgot that it was once respectable to see sea serpents, and for journalists to report on it. Ironically, it became respectable to see and report lake monsters.

In any case, a few do put in an appearance. This one was published in 2015, on pp 88 - 90 in the *Fortean Times* Magbook, *It Happened to Me!*, volume 1, which is a compendium of personal stories which its readers sent in. The witness, a Mr Nick Johnson told the story in 2002. One Sunday morning in 1987 he and a friend were fishing off Devil's Point, a quay on the Plymouth side of the narrowest point of the Tamar River, where the water was about 130 ft [*40 m*] deep and the current strong.

> A large head popped out of the water not 10 yards (9 m) from me, attached to a neck which rose out of the water by about 3 ft (90 cm). The head was covered with a fur-like, green-brown skin, had forwards facing, dark grey eyes (which looked directly at me) and was similar in shape to that of a large dog. It was not a seal, animals I have had much contact with as a diver. There were no ears but the top of the head was undulated with a high central ridge. It had a wide mouth and was obviously carnivorous from the shape of the powerful-looking jaw. Its forward facing eyes had fairly heavy brow ridges. From the size of the exposed neck and head, I would estimate the creature to be about twice the size of a horse.

He said that the creature was obviously sizing him up. It looked at him for about 15 seconds, submerged vertically, and came up again a bit closer for another 5 to 10 seconds. I would have liked to know how thick the neck was with respect to the head. And I wonder what it was.

Singapore, ? 1990s

This one was also published in the *Fortean Times* Magbook, *It Happened to Me!*, vol. 1, - in fact, right next to the Devon sea serpent story quoted above. This one was by S. J. Adams in 2001. I am not prepared to rule out a prosaic explanation, but the story should at least be publicised so that people can make up their own minds. I shall therefore repeat the entire account from page 90.

> I once saw a sea serpent in the well-used shipping lane at the back of Singapore Island. We were rounding Changi Point, only about half a mile (0.8 km) from the shore, entering the Johore Straits for the run up to Sembawang Dockyard.
>
> I was leaning against the rail, scanning the shore through a good pair of binoculars. There were two or three people on the beach looking out to sea, or at us, or whatever, when I spotted a grey-black, sinuous body, about one foot (30 cm) thick, with the conventional sea serpent humps, undulating through the water only about 150 yards (137 m) away in a flat sea. I saw no head, only a series of grey-black humps, and I watched it for at least half a minute, as we passed by at 11 knots. Then it quietly slipped below the surface and disappeared.
>
> By the time I looked around for someone else to confirm the sighting, the 'monster' had disappeared. I had an excellent view through binoculars; it was mid-morning and the bar had not yet opened. I am inclined to think that the people on the shore were also gazing at something strange.
>
> I wondered how I could fabricate such a 'monster' should I wish to deceive the guileless. The RAF still had a base at Changi, and I wouldn't put it past some of those jokers. But I never talk about it. After all, there are no such things as sea serpents, are there?

Off Japan, 1990s

In the first week of 2003, I was lying in bed at night, when I received a phone call from a Mr Mike Cleary, who was seeking specific information on unknown Australian animals for the husband of his niece in the UK. Unfortunately, I was not able to help him. Then he told me an incredible story.

He has been a diver for more than 35 years. About 10 years or so beforehand he was in a diving bell with a companion off the south-east coast of Japan, checking bottom sites for an oil rig. They were at a depth of 1,700 feet [> *500 metres*] when an unknown creature approached the bell and circled it.

It was about 25 foot [*7½ m*] long. It had no visible scales, and the skin changed colour in the light from the bell (which, I gather, is a common occurrence at this depth). It swam with horizontal undulations, and possessed just a single, elongated dorsal fin, extending down the body. I got the impression that it was like an eel's. He couldn't say much about the tail, but didn't think there was a tail fluke.

The head was like a sea horse's, the eyes like a cow's, and teeth like a barracuda's.

No constriction existed between the neck and body, but one ran into the other. However, 8 feet [*2½ m*] from the front was a pair a limbs, about 4 feet long. There was also a pair of hind limbs. I questioned him about this in particular, but he was emphatic that these were not fins, but webbed limbs.

What sort of creature could this be? The elongated dorsal fin and the horizontal undulations mean it had to be some sort of fish – but what? The obvious choice is some very large eel, or elongated shark – although none, to my knowledge, are of such a size. At a pinch, it might even have been an oarfish. However, it is the limbs that are the real problem. As you are no doubt aware, the vast majority of fish, the teleosts, possess rayed fins, which could hardly be mistaken for limbs, or even paddles. The largest fishes tend to be sharks, but their fins are also hard to mistake for limbs, and most people – especially divers – would be familiar with them. But there once existed a vast array of lobe-finned fish, of which only a few relic species are now known to exist. One is the Dipnoi, or lungfishes, and the other is the Crossopterygians, whose lobed fins evolved into the legs which all land vertebrates now walk on. However, except for the two species of coelacanth, they all went extinct about the same time as the dinosaurs, and were on the way out for a long, long time before that.

So, if Mr Cleary's perception and memory were accurate, something very strange was swimming around off the coast of Japan.

Chapter 8

Lakes and Swamps

As I said before, although a great reluctance has developed nowadays for people reporting sea serpent sightings and newspapers covering it, the same cannot be said for lake monsters. In the past, it was the reverse. But here are a couple which have probably avoided the eye of most researchers.

Lake Minnetonka, USA, 1887 and 191
This is a story 'bout Minnie the Monster.
Sorry! I couldn't resist that. Readers of my generation will recognize a parody of a popular song, *Minnie the Moocher*. In any case, one of the strange mysteries of cryptozoology is how certain lakes gain a reputation as being the habitation of a monster, which only reappears at very long intervals. Take Lake Minnetonka, Minnesota, for example. It's adjacent to the huge St. Paul-Minneapolis conurbium, for heaven's sake! - hardly the site you'd predict for a self-respecting monster to hide. I know the Wikipedia says that a big sturgeon called Lou is supposed to dwell there, but this is small fry. I'm talking about Minnie, and she's big!

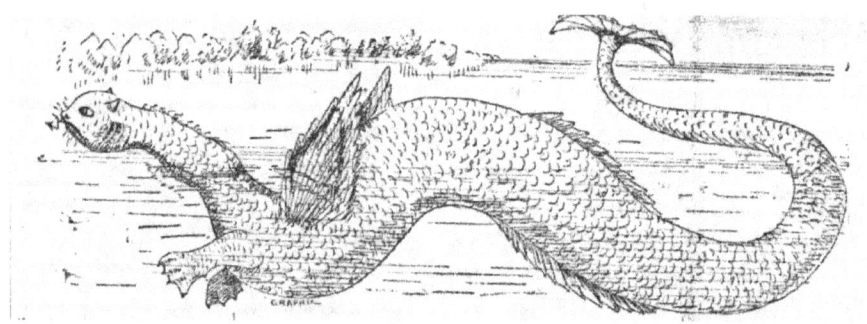

The above is a drawing of Minnie. It was taken from the *St Paul Daily Globe* of Tuesday morning, May 31, 1887. So, here is the story.

HERE'S A SNAKE,
Unless the Eyes of Reliable Witnesses Have Been Deceived.
Dwellers at Lake Minnetonka Say They Saw a Monster
With a Terrible Head and Fins That Moved Like Wings.
A Great Leviathan Sporting the Big Water This Year.

On Several Occasions He Has Lashed the Lake to Foam.
The General Topic of Conversation at the Lake.

Snake stories are always disbelieved, save by those who actually come in contact with representative reptiles, and the writer of a communication to the GLOBE doubtless felt the same way, for he not only signed his own name, but gave information which yielded good fruit in an investigation. In the mail that reached the GLOBE Saturday was the following missive: :

Wayzata, Minn., May 27, 1887.

DAILY GLOBE, St.Paul, Minn, Inclosed we send you a picture of the lake serpent as seen by a party of fishermen in Lake Minnetonka on the evening of the 23th inst. The immensity of this strange amphibious animal frightened the party from the lake. We have the names of about fifteen persons that have seen this serpent.

MARTIN V. HENRY, one of the party.

The picture referred to was a pen and ink sketch of a monster not unlike a gigantic snake with two fore paws, or flippers, and a forked tail similar to that usually delineated in representations of the prince of darkness, and in the middle of the body were two gigantic fins, which at first glance, looked like wings. The dimensions of the monster were given as thirty feet [*9 metres*] in length and about as large round the belly as a full-grown man.

Armed with the document and picture a representative of the GLOBE set out for Wayzata and was fortunate enough to find, upon leaving the cars at the depot, an eye-witness of the maneuvers of the monster in the person of A. P. Dickey, at present engaged in building a bridge across Shafer's narrows, a point opposite to Wayzata on the west. According to his statement, several days ago while at work on the bridge with Messrs. George McLean and Jacob Snow, the narrator saw what at first looked like a log moving down Lake Minnetonka from Cedar point.

"You will all say this is another snake story," he said, looking around at a group of open-mouthed rustics who had been attracted by the advent of a stranger and a note book, "but what I am telling you I saw with my own eyes. My companions on the bridge did not see the thing, whatever it was, until I directed their attention to it, and finally it came within 200 yards of us. Raising its head several feet, for by this time it gave unmistakable signs of life, the monster began to thresh the water violently, and that

IT WAS THIRTY-FIVE FEET LONG. [*10.6 m*]

"Closer inspection made it look like the head and foreshoulders of an alligator, for it came within six or seven rods [*33 to 38½ yards*] of the bridge before it finally dived out of sight, and we did not see it rise again."

All the details were listened to with the deepest attention by the crowd now gathered about the story-teller, who was an honest-looking, well-dressed fellow, and too much in earnest with his subject to give the faintest suspicion of intoxication. Then it occurred to one of the group that a lady named Thurston had also said something about seeing a monster last Saturday in the lake. A walk of a few minutes over a pleasant stretch of sward soon brought the GLOBE's representative to a brown painted cottage on a bluff overlooking the lake and surrounding country.

In response to a knock at the cottage door Farmer Thurston appeared and ushered his visitor into the parlor, where the mistress of the house was sitting.

"Last Saturday afternoon, the 21st of May," she began, "just after the shower, I had occasion to go to the edge of the bluff to look after my children. I did not see them for a few minutes, and stopped to listen for sounds from them. While thus waiting my attention was attracted to what at first seemed to be the struggles of a man drowning in the lake, a few feet beneath where I stood: The water was lashed to a foam by something, and while I looked more intently, I saw it come up out of the water with head erect, several feet. What I thought were two flippers were moving continually, and I at once saw that it was a snake, or big water monster of some kind. It turned over several times, showing it to be very long, and kept me fascinated by the unusual spectacle for several minutes. There were no boats or fishermen near by, the storm of a few minutes before having driven them all ashore, and I could not call any one to see the monster. As soon as my husband returned from his work I related to him what had occurred, but he advised me to keep it quiet; as people would think it was only an optical illusion. I saw the reptile, or what ever you might term it, as plainly as I ever saw anything in my life, and it was fully thirty feet long and as large round as a man's body. Mrs. Gallagher, up on the hill, has also seen the monster, but she was closer to it than myself and saw its outlines better."

ANOTHER EYE WITNESS.

A trudge up a steep declivity to the Arlington hotel, which covers a high bluff on the Huntington estate, was rewarded by a sight of Mrs. Mary A. Gallagher, who is in charge of the premises, and she promptly acceded to a request to tell what she knew about the lake terror.

"One afternoon I was down on the wharf below the boat house with my children," said Mrs. Gallagher, "and they were amusing themselves wading and splashing about in the water. Several other children were with them engaged in the same manner, when something caused me to look up suddenly. What at first looked to me like a colored man floating in shore caused me to shout to the children, 'There is a man swimming toward you; come ashore.' My first impression was that some one had jumped off the steamer, intending to have a swim, and as I did not want to see him land I naturally called to my children. At first they misunderstood me, but eventually they saw the object which was now several feet above the surface of the water, and they screamed with terror. Their shouts evidently scared the serpent, for such it appeared to be, having a large flat head with what seemed to be bushy black hair, and it lashed the water violently and disappeared. I was very much frightened, and a party of fishermen, attracted by our noise on shore, looked up from their lines, and as they did so the monster passed their boat. One of the ladies of the party screamed in terror, but the snake swam by without attempting to injure anybody. At least that was what they told me when they came ashore. I could have at one time hit the serpent with a long stick, for it was not further from me than the length of this room, about twenty-five feet [7½ m], and I distinguished its eyes to be light. Its belly glistened as it turned over apparently, and the color seemed about that of a catfish, and I should say it was between twenty-five and thirty feet long. This was my first, and I hope, it will be my last, look at the serpent or whatever it was."

Conversations with numerous denizens of the locality elicited the information that the people whose statements have been given were thoroughly reliable, but they had refrained from repeating their experiences, because they feared the ridicule that might follow.

But the adventure of the fishing party the evening of the 23d inst. had recalled the incident more vividly than ever, and now it was the principal topic of discussion whenever a party assembled for the evening. No fishing party at Wayzata considers its make-up complete unless there is a heavily-loaded gun in the bow of the boat ready for use, and many of the fishermen have additional security in the form of revolvers worn around their waists.

It seems our friend made further appearances in the following decade, but I have no information on that period. But when she came back in 1914, the news even reached Australia. The earliest report appeared in the *Journal* (Adelaide) of Saturday 20 June 1914, on page 7.

AN AMERICAN SEA SERPENT

After an absence of 17 years, Lake Minnetonka's sea serpent has reappeared on the waters of Lake Minnetonka, according to numerous residents of the village of Wayzata. At least, it is generally believed that it was the same old serpent, although some admit that it might be the son or grandson of the original marine monster that invaded the otherwise peaceful lake about 1895. On the present visit it appeared twice one day and once the next. It was seen by nearly a dozen persons, including the postmaster, the telephone operator, business men and others.

All agreed that the " thing"'
Was from 12 to 20 ft. [*3.6 to 6 metres*] long.
Swam at either 30 or 60 miles an hour [*48 to 96 kph*].
Was at one time within 40 rods [*220 yards*] of the shore.
Had a snakelike head "as big as a bucket."
Had four or five black fins that kept rising and falling.
Beat the water into a froth when it swam.

The uncanny-looking visitor, according to those who saw it, first appeared at 11.30 a.m. 40 rods off Wayzata dock. It turned and swam rapidly across the bay in the direction of Breezy Point, turned and started back, observers said. In the afternoon the strange creature was seen, this time far out in the bay and only for a few seconds, and then it seemed to sink back into the lake. Its third and last appearance was at 7.15 a.m. the next day, when a lone telephone girl, completing her night's work, glanced out of the window of the telephone exchange, and saw the same old "serpent" speeding through the water toward Breezy Point at a great rate.

Frederick Rodner, of the firm of Bradshaw & Rodner, was the fortunate individual who first spotted the visitor when it made its initial appearance.

"The 'thing' was about 40 rods off shore at that time," he said. "The hour was just 11.30 a.m., and E. G. Braden, the postmaster, was with me.

"When I saw it first it apparently had just turned toward Breezy Point, and was speeding away in that direction. I called to Mr. Braden and pointed it out to him. Others soon gathered. We watched the thing cross the bay at lightning speed. It was lost to sight somewhere near Cedar Point.

"The 'thing' swam 60 miles an hour. It was about 16 ft. [*4.9 metres*] long and had five black fins that kept rising and falling above the water as it swam. It left a trail of froth as it hurried along, and could be still be seen when a mile away. It had a head like a snake. A man can't remember everything at a time of great excitement like that."

The postmaster (E. G. Braden) confirmed Mr. Rodner's story except as to dimensions and speed. "The 'thing" was from 12 to 20 ft. long," he said, "and moved through the water at the rate of at least 30 miles an hour. I won't try to say what it was. It looked like a serpent ought to look. It was black, and had a head like a snake. At least that was the way it looked to us. At first I thought it was four ducks swimming in single file. But there are no ducks that I ever heard of that can swim 30 miles an hour, and be seen more than a mile away."

In the little telephone exchange at Wayzata, the day operators saw the "thing."

"It was terrible looking," said Miss Alexa Shaw, one of the operators. "It was half way across the bay when I first saw it. It had a head as big as a bucket, and shaped exactly like a snake's head. It kept this head raised out of the water all the time rubbering. It swam very fast, and disappeared near Cedar Point. It had four black fins."

Miss Eulalia Bleakley, another telephone operator, said she saw the strange visitor.

What can one say? In the 1880s, journalistic practical jokes were not unknown, but I doubt if they would be audacious enough to quote several inhabitants of a small village, where everybody knew everybody else. Also, by 1914, the age of journalistic hoaxes was more or less past. Therefore, we must conclude that several people was conspiring to lie through their teeth, or they really saw something odd. But what?

A lake monster which appears only at very long intervals cannot be an air breather. Just imagine if, for example, a whale or dolphin somehow got into the lake. Considering how often it had to come up for air, how long do you think it would be before everybody had seen it, it had been identified and acquired a name, and the press were providing daily up-dates on its movements?

In addition, in respect to these reports, no snake has fins on its side, and no eel is anywhere as long as described, nor does any possess a lobed tail. Also, I cannot help feeling that the 1914 Minnie was a different one to the 1887 one, due to the presence of four or five fins. Note, too, that their rising up and down indicates that the animal was a mammal. I am also interested to know how anybody can make an estimate as precise as 16 feet at a distance of 220 yards. And even if the speed was greatly exaggerated, it does seem a little on the high side for a known animal..

"Curiouser and curiouser," said Alice.

Crescent Lake, Oregon, USA, 1910

"Cressie" is the monster of Crescent Lake - in Labrador. But North America alone possesses nine other lakes with that name. (There are also sixteen Round Lakes and five Square Lakes. Do you get the impression that some name-givers are not terribly imaginative?) Do any of the other nine possess a Cressie? Here is a story which apparently went around the world, because I found it in a newspaper in Adelaide, South Australia: the *Express and Telegraph* of Saturday 5 November 1910, at page 4.

SEA SERPENT.
Seen in Summer Residence in Crescent Lake

> The remarkable fact of the non-appearance of the sea serpent this season has not hitherto been accounted for, but his absence from observation is now explained by a dispatch from Eugene, Oregon, quoted by the New York correspondent of the "Liverpool Courier."

"The creature," he says, "has left the ocean for the time being and taken up its residence in Crescent Lake, a huge lake in the Cascade Mountains, 100 miles east of Eugene.

"This is according to the statement of a Dr. Wood and Mr. E. L. Lampson, prominent residents of Portland, Oregon, who have just returned from a hunting and fishing trip. They declare that, while fishing in the lake they saw a huge creature with a head as big as an ox and a long body and tail. The monster rose from the centre of the lake and surveyed them with large saucer eyes, but did not attempt at [*sic*] attack them.

"Naturally such sportsmen did not let the opportunity pass without trying to 'bag' the creature, and they fired 25 shots at it from their rifles. Either the bullets missed - be it noted that the hunters are honest enough to admit the possibility - or else the creature possesses an armor-plate hide, for it just glanced at them again and leisurely glided below the surface.

"Indians confirm the hunters' story and say that they have seen the creature before, although it only appears at very rare intervals. They have a tradition that it has been there for all time, but if this is so the poor old sea serpent has been able to establish an alibi.

"The American newspapers scout such an idea, and declare that Crescent Lake is obviously a summer residence visited at intervals by our old friend the scaly monster of the deep. For most conclusive proof of all, Oregon is practically a prohibition state."

So there you have it. If Crescent Lake, Oregon has a "Cressie", I have not heard any further mention of it. As you can see, the account lacks a few relevant details - like distance, size, colour, general shape etc, even whether they were on shore or in boat at the time. (If they were in a boat, why were they carrying rifles?) As I explained before, the rarity of lake monster sightings indicates that, whatever they are, they don't breathe air, and the Indians' tradition that it appears only at very rare intervals would appear to support it. Despite the reference to "large saucer eyes", I wonder if the animal were not a sturgeon which had somehow found its way into the lake, and had grown to a great age and a great size. The white sturgeon, *Acipenser transmontanus* is mature at 160 cm or 5 ft 4 inches, but has been recorded as reaching 610 cm, or 20 feet.

A correspondent suggested to me that it might have been some sort of huge tortoise, but if so, it would need to come ashore to lay eggs.

Everglades, Florida, 1901

I know this isn't actually a *sea* serpent nor, strictly speaking, a lake monster. However, it is a serpent, and the incident is little known, and is too good to be allowed to pass. Besides, this is my book; I'll put in it whatever I like.

Since the 1980s, Florida has been infested with an invasive species, the Burmese python (*Python bivattatus*), and early this century someone discovered a green anaconda, the largest snake in the world, there[4]. But what about the old days? Here is a report which appears to have gone around the world, because I discovered it in an Australian newspaper, the *Maitland Daily Mercury* (NSW), Tuesday 15 April 1902, on page 5, but it seems to originally appeared in the *New York Times* of 30 November 1901.

A Strange Monster

An enormous reptile, more like the extinct brontosaurus or fabled sea serpent than any living creature, has (says a Jacksonville correspondent of the New York *Times*) recently been killed by a hunter in the lower Everglades. It has for one hundred years not only been a tradition among the Seminole Indians, who inhabited the borders of Lake Okeechobee, but it is stated as a fact within the knowledge of some of the Indians now living that an immense serpent made its home in the Everglades, and has carried off at least two Indians. The Indians reported the animal to be snakelike in appearance, with ears like a deer; that it had only been seen in the Everglades, and that it was very wild. They said that when it travelled it frequently stopped, raised its head high above the sawgrass to take a view of its surroundings, to discover enemies, or locate victims. If frightened, the Indians asserted that it glided off at immense speed. These stories have kept the venturesome hunter and trapper on his guard and in a state of more or less anxiety, notwithstanding they did not give credence to these Indian stories.

Recently Buster Ferrel, one of the boldest and most noted hunters of Okeechobee, and who for twenty years has made the

---

[4] Jackson Landers, 15 Nov. 2013, 'The largest snake in the world has invaded the United States',
https://slate.com/technology/2013/11/green-anacondas-in-the-everglades-the-largest-snake-in-the-world-has-invaded-the-united-states.html

border of the lake and the Everglades his home, on one of his periodical expeditions noted what he supposed to be the pathway of an immense alligator. For several days he visited the locality with the hope of killing the saurian, but was unsuccessful in finding him. His pride as a hunter was piqued, and his desire to obtain the hide of what he felt sure to be one of the largest alligators ever seen in this section, where alligators are noted for their immense size, grew daily. He studied some plan to outwit it. A large cypress stood near its pathway, and he concluded to climb the tree and take a stand for his game. He accordingly took his position in the tree. Nothing appeared. He was becoming discouraged, but determined to give one more day to the effort. On the third day, before he had been on his perch an hour, he saw what looked to him like an immense serpent gliding along the supposed alligator track. He estimated it to be anywhere from 25ft. to 30ft. [*7½ to 9 metres*] long, and fully 10in. to 12in. [*25 to 30 cm*] in diameter where the head joined the body, and as large around as a barrel 10ft. [*3 metres*] further back.

The creature stopped within easy range of his gun and raised high its head. As it did Ferrel shot at its head. Taken by surprise, the serpent dashed into the marsh at tremendous speed, while Ferrel kept up firing until he had emptied the magazine of his rifle. About four days afterwards he ventured back into the neighbourhood, and about a mile from where he first saw the monster he saw a large flock of buzzards, and went to see what they were after, and there he found the creature dead and its body so badly torn by the buzzards that it was impossible to save the skin. He, however, secured the head, and has it now in his home on the Kissimmee River. It is truly a frightful looking object, fully 10in. [*25 cm*] from jaw to jaw, and ugly, razorlike teeth. He described the animal as dark coloured on its back and a dingy white beneath, with feelers around its mouth similar to catfish. He has gone back into the swamp with the intention of obtaining the skeleton and bringing it back, after which he will send it to the Smithsonian Institution in Washington.

That last sentence is telling. Was the skeleton ever really forwarded to the Smithsonian? I did a web search, but was unable to discover any further reference to the snake, the skeleton, or the hunter. Perhaps he made it up. Pity.

Help Wanted

You will probably never see a sea serpent, but if you do, I would like to hear from you.

More to the point, you might be rummaging through old newspaper files, or perhaps your local paper has been digitalised, and you are able to run a search for the words, "sea serpent". You might discover reports which have been overlooked by the previous researchers - or perhaps the original article of the cases recorded here. Indeed, you may even discover more recent sea serpent sightings. If so, I would like to hear from you.

I operate a blog on mystery animals - only occasionally updated these days - at https://malcolmscryptids.blogspot.com/ . At the top you will see a button labeled "How to report a sighting". Or, if you wish, you can go straight to that page:
https://malcolmscryptids.blogspot.com/p/how-to-report-sighting.html .

This will lead you to my e-mail address. It will also outline the details which would be useful if you yourself are a witness.

INDEX

Abdurahim, 82
Adams, S.J., 111
Aden, 107-108
Akaroa, 44
Allen, Captain, 13
alligators, 15
Alpha, 8
American (steamer), 50
Anchoria (steamship), 25
Andrews, Mr., 11
Arawata, 39
Arbroath, 51
Aro, S. S., 92, 107
Australian National Library, 2
Avery, Edgar, 36
Australian Sea Serpents, 2, 75
Ballard, Captain J., 55 ff
Baltic Sea, 98
Banks Peninsula, 43
Bartlett, Captain W. H., 60, 67-68, 72
Bartram, A., W. & R., 43
Batu Islands, 54
Benson, Captain H, B., 9-10
Bienvenu, 29
Bleakley, Miss Eulalia, 118
Bonifacino, 79
Borneo, 90, 96
Braden, E. G., 118
Brazil, 80
British Channel, 9
Brothers Lighthouse, 41
Bryan, Thomas, 42
Bullen, Frank T., 87 ff
Bullock, Samuel, 28
Burmese python, 121
Cable Bay, 46

Cadborosaurus, 36, 48, 77, 83, 85, 103, 105
Cape Bridge, 103
Cargill, Joe, 52
Case for the Sea-Serpent, The, 1
Christian, H. C., 45
Chronicles of the Strange and Uncanny in Florida, 49
Cikobia, 21-22
Clark, Captain, 91
Cleary, Mike, 111
Connecticut, 28
Connecticut (steamer), 5
Cook Strait, 41, 44-45
Coral (schooner), 28
Corbiere Lighthouse, 106
Cornfield Point Lightship, 28
Craig Gowan (steam trawler), 55 ff
Craven, Captain J., 93
Crescent Lake, 119-120
crocodiles, 15
Crossopterygians, 112
cryptozoology, 3, 5
Cryptozoology, 85
Cumberhall (steamship), 80
Cutting (mate), 23
Daedalus, 7, 10
Darius (steamship), 53
Dartmouth, 108
Davis, Nurse, 101
Delory, Captain, 35-36
Despatch (barque), 10
Devon, 110
Dickey, A. P., 114
Diligence (fishing boat), 51
dolphin, 54
Doney, George W., 48

Doyle, James, 24
Dublin, 10
Dunottar Castle, 89
Easler, Alexander, 76
Eberhart, George M., 30
Edwards, Captain, 8
Elizabeth, Cape, 18
Emery, Prof. Stephen, 28
Estrella (barque), 36
Evans, James, 42
Everglades, 121
Falmouth (barquentine), 23
Ferrel, Buster, 121-122
Fiji, 21-22, 82, 101
Flavell, Charles, 42
Florida, 121
von Forstner, Baron, 07
Fortean Times, 110-111
Foster, C. W., 31
Fotheringay, Ethelbert G., 99
Fraserburgh, 55
Fuller, E. G., 24
Galápagos Islands, 37
Gallagher, Mary A., 116
George, Constable, 90
Georgetown, 30
Gibsons, 105
Glanrafon (bark), 17
Godson, Mr., 83
Gould, R. T., 1
Grantully Castle, 107
Graves, the, 5
Great Sea-Serpent, The, 1, 30
Greg, R. P., 30
Grey, Joseph Ostens, 59 ff
Griffith, Sir William Bradford, 92
Griffiths, Elias, 60, 68, 71-72
Groves, Harold, 108
Gulf of Aneud (steamer), 54

Gulf Stream, 13, 17
Hart, John, 20
Hastborough Sand, 85
Hatteras, Cape, 59 ff
Heathdene (steamer), 77
Hebda, Andrew J., 81
Hecate Strait, 103
Henry, Martin V., 114
Henry Buck (tug), 30- 32
Heuvelmans, Dr. Bernard, 1-3, 8,
    19, 26, 30, 44, 49, 59, 69-71,
    75, 80, 83, 86, 96,103, 107, 1
    110
Hilary, H.M.S., 38
Hogan, Mr., 10
Hong Kong, 75 ff
Horawell, W. J., 17
House, Captain, 103
Howell, Mr., 39
Howell, C. & H., 43
Hubbard, Captain, 32
Hurd, James, 28
Iberian (steamer), 97
ichthyosaurs, 36
In the Wake of the Sea-Serpents,
    1, 30, 44, 59, 103
Inconstant (barque), 7
Instone, Mr., 54
International Society of
    Cryptozoology, 85
It Happened to Me!, 110-111
Jackson, Captain, 28
Jacksonville Beach, 49
Japan, 112
Java Sea, 102
Jenkins, Greg, 49
Jersey, 106
Jessie Rosaline (schooner), 31
Johnson, Nick, 110

Kawharu, Mrs. E. E., 46
Kent, 16
Kermadecs, 87 ff
Kirkwall, 26
Koopman, Captain J., 85-86
Kuster, V., 76
Labuan (steamer), 90
Lampson, Mr. E. L., 120
Largs Bay, S. S., 108
Las Palmas, 92
LeBlond, Paul, 85
Lelean, Rev., 101
Lincoln, 43
Linwood, 27-28
long necked seals, 2
long necked sea serpent, 40, 41, 78, 86, 109
Los Angeles, 104
Lucille, 9
Lum-Lum (steamer), 99
Lungting (launch), 75 ff
Madora (schooner), 40
Maggie Comb (trawler), 55
Mahinapua (steamer), 42
Maine, 18
Mangawhare, 42
Mary Lane (sloop), 35
Massachusetts, 5, 26-27
Maude, Cyril, 108
McLean, George, 114
M'Donald, Matthew, 24
Mereddlo (steamship), 91
metric measurements, 3
Meli, 82
Milburn, Mrs. & Miss, 77
Minnetonka, Lake, 113 ff
Mohammed Singh, 96
Montrose, 84
M'Taggart, Captain, 16

Munro, Captain, 44
Murray, Neal S., 91
Mysterious Creatures: A Guide to Cryptozoology, 30
Nahant, 5
Naish, Darren, 70
Nautilus (bark), 37
Neeson, Mr., 54
Nelson, 46
Nelson, Captain, 19
New England, 5, 11, 18, 23. 26, 28
New England (steamer), 78
New Jersey, 11
New York, 22
New Zealand, 39 ff
Newfoundland, 25, 50
Norfolk, 85
North Cape, 44
Northumberland, 55
Okeechobee, Lake, 121-122
Oliver, Captain, 18
Onward (brigantine), 16
Oregon, 36
Orkneys, 26, 84
Osborne, David, 83
Oudemans, Dr. A. C., 1, 30
Owen, Prof., 7, 88
Owen, A. W., 10
Pablo Beach, 49
Padang, 54
Paijmans, Theo, 33
Park, George, 8
Park, William, 105
Parkhurst, Clinton Gordon, 104
Parnell, Frederic, 105
Parsons, Mr., 29
Pelorus Sound, 45
Pepin Island, 46

Pernambuco, 86
Peters, First Officer, 50
Pheonix, H.M.S., 54
Planter (schooner), 32
plesiosaurs, 2, 36, 89
Poole, Calvin W., 27
Pooley, Captain Ric, 92
Poor Knights Islands, 40
Portland, 18
Prince Edward Island, 24
Puget Sound, 48
Pulau Bojo light, 54
Pulau Tanahbala, 54
Punch, Cyril, 92
Putnam, G. B., 27
Racher, W. Messenger, 10
Radnorshire (steamer), 21
Rattray Head Lighthouse, 55-56
ray, 8
Reay, S. G., 10
Recife, 86
Redhead, 51
Redlap Cove, 108
Rhode Island, 35
Rock Island, 20-21
Rodner, Frederick, 117
Ryukyu chain, 20-21
Sacramento, 19
Sardinia, 79
Scottish Bride, 13
Sea Serpent Chronicles, The, 81
Sehome, 48
Sharp, J., 17
Shaw, Miss Alexa, 118
Shepherd, David, 52
Sherman, Captain, 28
Siberut, 54
Sierra Leone, 92
Singapore, 111

Smith, D., 106
Smith, Hugh, 52
Smith, Major O., 98
Smith, Robert, 52
Smith, Captain William F., 37
Smythe, J. J. R., 107
Snow, Jacob, 114
South Carolina, 30
Springs, Captain A. A., 30, 31
Staveley, Captain W. E., 80
Stephen's Island, 41
Stonehaven, 89
Strait of Georgia, 105
Strathardle (cargo steamer), 93-94
Strathspey (tramp steamer), 95
Street, H. B., 48
Sumatra, 53-54
Suncombe, S. E., 10
Swakopmund, 99
Tamar River, 110
Tasman Bay, 45
Taveuni, 101
Taviuni (steamer), 87 ff
Terawa, 45
Thompson, Mr., 8
Thurston, Mrs., 115
Timaru, 40
Trove, 2
Tregurtha, 41
Tresco (steamer), 59 ff
U28, 70, 97
Umpqua River, 36
Umtentweni, 109
Utting, Matthew A., 97
Van der Woof, Captain Johann, 99-100
Vancouver Island, 78
Venice, 104
Vernon, William, 7

Victoria, B. C., 83
Wairoa, 42
Walsh, Mr., 10
Watkins, Charles, 7
Watt, J., 56
Wayzata, 114 ff
Webster, Mr., 29
Wegener, Mr., 97
Western Islands, 17
whale, 18
    blue, 55
    humpback, 30, 88
White, W., 17
Wide World Magazine, The, 59, 69, 71
Wilson, Mr., 54
Winslow Morse (schooner), 18
Wolfe, F., 75-76
Wood, Dr., 120
Wright, Sergeant, 90
Wright Sound, 103

Other Books by the Author

The following books are all in print, and available from Amazon.

Bunyips and Bigfoots. up-dated second edition. The classic survey of Australia's mystery animals: bunyips, sea serpents, the North Queensland tiger, Tasmanian tigers on the mainland, pumas and black panthers, yowies, and others - all fully documented. Originally published in 1996, it was republished and up-dated in 2021.

Australian Sea Serpents. The companion book to this one. Sea serpents really exist, and have been visiting Australian shores for a long time. I have now trawled through dozens to digitalised newspapers to produce what I expect will also become the definitive work on the subject. More than half of the case histories are new - that is to say, they have never been published in book form by other researchers.

The Truth About Bunyips. Every Australian has heard about bunyips, but no-one knows what they are supposed to look like. Based on a huge number of recently digitalised old documents and newspapers, this short book should be the definitive work on the subject.

Forgotten Bigfoots Around the World. Here I present a series of articles - mostly my own translations - of reports from difficult-to-access journals about bigfoot type animals from the Caucasus, the Himalayan region of Pakistan, as well as Latin America, Africa, and even Europe.

The Stranger from the Stars. A science fiction novel about a group of hikers who rescue an injured alien from a crashed flying saucer. Having followed the UFO scene for more than 50 years, I have ensured that the story is "realistic", in that all the phenomena described have been reported many times in the literature.

Savages and Saints by Leon and Theophila Philippi - but ghost written by yours truly. This is the story of my parents-in-law: a farm boy from Nebraska, and a pastor's daughter from the Eyre Peninsula of South Australia, who were thrown together under unusual circumstances, married after a whirlwind courtship, and set out for New Guinea as missionaries. The sort of experiences they went through are beyond the imagination of the present generation.

The Gospels: Harmonized and Annotated (two volumes): Here I present the four gospels together in chronological order, with parallel texts side by side, and with a discussion of the situation in first century Israel in which they were embedded.

Trials of a Tourist. I've been an international tourist for most of my adult life, visiting remote places even millionaires haven't seen. So now I have written a humorous account of the quirky things I have experienced, as well as the "plot against tourists", by which the world conspires to make travelling as inconvenient as possible.

Apparitions: tulpas, ghosts, fairies, and even stranger things. This is not an ordinary book on ghosts, but coves a whole range of paranormal apparitions: tulpas and other creations of the mind, ghosts, fairies, and things which are more bizarre - but all fully documented.

A Zoologist Looks at Science Fiction. H. G. Wells said that the essence of science fiction was the suspension of disbelief. As a zoologist, I have a bit more difficulty than most. In this short book I examine the mistakes made by science fiction writers in their creations of monsters and aliens, not to mention robots.

The Repat Racket. An insider's report on Veterans' Affairs. Originally published in 2010, and now republished, it reveals the way in which good intentions for compensating ex-servicemen resulted in a grotesque legal system open to enormous abuses.